THE MORAL CHALLENGE OF
DANGEROUS CLIMATE CHANGE

This book examines the threat climate change poses to the projects of poverty eradication, sustainable development, and biodiversity preservation. It offers a careful discussion of the values that support these projects and a critical evaluation of the normative bases of climate change policy. This book regards climate change policy as a public problem on which normative philosophy can shed light. It assumes that the development of policy should be based on values regarding what is important to respect, preserve, and protect. What sort of climate change policy do we owe the poor of the world who are particularly vulnerable to climate change? Why should our generation take on the burden of mitigating climate change that is caused, in no small part, by emissions from people now dead? What value is lost when natural species go extinct, as they may well do en masse because of climate change? This book presents a broad and inclusive discussion of climate change policy, relevant to those with interests in public policy, development studies, environmental studies, political theory, and moral and political philosophy.

Darrel Moellendorf is Professor of International Political Theory and Professor of Philosophy at Johann Wolfgang Goethe Universität Frankfurt am Main. He is the author of *Cosmopolitan Justice* (2002) and *Global Inequality Matters* (2009). He coedited *Jurisprudence* (2004, with Christopher J. Roederer), *Current Debates in Global Justice* (2005, with Gillian Brock), *Global Justice: Seminal Essays* (2008, with Thomas Pogge), and *The Handbook of Global Ethics* (2014, with Heather Widdows). He has been a member of the School of Social Sciences at the Institute for Advanced Study (Princeton), a recipient of DAAD and NEH Fellowships, and a Senior Fellow at Justitia Amplificata at Goethe Unviersität Frankfurt and the Forschungskolleg Humanwissenschaften, Bad Homburg.

The Moral Challenge of Dangerous Climate Change

VALUES, POVERTY, AND POLICY

DARREL MOELLENDORF

Johann Wolfgang Goethe Universität
Frankfurt am Main

CAMBRIDGE
UNIVERSITY PRESS

CAMBRIDGE
UNIVERSITY PRESS

32 Avenue of the Americas, New York, NY 10013-2473, USA

Cambridge University Press is part of the University of Cambridge.

It furthers the University's mission by disseminating knowledge in the pursuit of education, learning, and research at the highest international levels of excellence.

www.cambridge.org
Information on this title: www.cambridge.org/9781107678507

© Darrel Moellendorf 2014

First published 2014

Printed in the United States of America

A catalog record for this publication is available from the British Library.

Library of Congress Cataloging in Publication Data
Moellendorf, Darrel.
The moral challange of dangerous climate change : values, poverty, and policy /
Darrel Moellendorf, Johann Wolfgang Goethe Universität Frankfurt am Main.
    pages cm.
Includes bibliographical references and index.
ISBN 978-1-107-01730-6 (hardback) – ISBN 978-1-107-67850-7 (pbk.)
1. Climate change – Moral and ethical aspects.   2. Environmental ethics.   I. Title.
GE42.M64   2014
179′.1–dc23   2013038844

ISBN 978-1-107-01730-6 Hardback
ISBN 978-1-107-67850-7 Paperback

For Bonnie and Marino with whom I have shared sunrises and snowstorms in the Colorado Desert

I open the books on Right and on ethics; I listen to the professors and jurists; and, my mind full of their seductive doctrines, I admire the peace and justice established by the civil order; I bless the wisdom of our political institutions and, knowing myself a citizen, cease to lament I am a man. Thoroughly instructed as to my duties and my happiness, I close the book, step out of the lecture room, and look around me. I see wretched nations groaning beneath a yoke of iron. I see mankind ground down by a handful of oppressors. I see a famished mob, worn down by sufferings and famine, while the rich drink the blood and tears of their victims at their ease.

– Jean Jacques Rousseau

Contents

Acknowledgments

This book was written mostly while I was a member of the Philosophy Department at San Diego State University (SDSU). Colleagues who work at public universities devoted primarily to undergraduate teaching know how hard it is to get the time and support necessary to read, reflect, and write. I have been very lucky in this regard.

This book would never have been written without the support of several institutions and people within them. Due to the efforts of Marius Vermaak, the Philosophy Department and the Office of International Education at Rhodes University in Grahamstown, South Africa, hosted me for several weeks in 2006 when I first started to work on climate change. The Deutsche Akademische Austauch Dienst provided me with a fellowship that allowed me to spend the summer of 2008, at the invitation of Stefan Gosepath, at the Institut für Interkulterelle und Internationale Studien (InIIs) at Universität Bremen. InIIs also generously found work space for me every summer between 2008 and 2012. Chapter 5 was first drafted there in 2010. SDSU has supported me with two leaves, one in 2008–2009 and the other in 2012–2013. The College of Arts and Letters at SDSU provided me with the Johnson Critical Thinking Grant in the fall of 2011 that relieved me of some of my teaching duties. The School of Social Science at the Institute for Advanced Study offered a membership and the Friends of the Institute for Advanced Studies supported me in 2008–2009, during which time the first and fourth chapters of this book were starting to form. Once again due to Stefan Gosepath's efforts, Projekt Justitia Amplificata at Goethe Universität, Frankfurt, invited me to be a Visiting Fellow and to reside at the Forschungskolleg Humanwissenschaften in Bad Homburg during the summers of 2010 and 2011 and to be a Senior Fellow, again with residence at the Forschungskolleg during the year of 2012–2013, when this book was finished. At both the Institute for Advanced Study and the Forschungskolleg Humanwissenschaften I was aided immensely by very

helpful staff, especially by Kirstie Venanzi and Donne Petito at the Institute for
Advanced Study and Andreas Reichardt, Ingrid Rudolph, and Beate Sutterlüty
at the Forschungskolleg. Andreas saved me a great deal of time by getting books
for me from the library in Frankfurt. This book would probably not have been
written were it not for the support of these institutions and the people who
make a difference within them. I am deeply grateful for their help.

More people than I can remember, I am afraid, have spoken to me about
this project. My thinking about it has improved as a result. For the con-
versations that I have had, and the comments that I have received, I am
thankful to all of you. Here is a partial list to those who have helped in
this way, with apologies for my failing memory to many people who should
also be on the list: Danielle Allen, Robin Attfield, Paul Baer, Gillian Brock,
Daniel Callies, Simon Caney, Marion Carlson, Thomas Christiano, Diarmuid
Costello, Thomas P. Crocker, Julian Culp, Timothy Diette, Freeman Dyson,
Maria Paola Ferretti, Rainer Forst, Josef Früchtl, Stephen M. Gardiner, Stefan
Gosepath, Carol Gould, Andrew Greetis, Klaus Günther, Nicole Hassoun,
Meghan Helsel, Michael Howard, Aaron James, Dale Jamieson, Anja Karnein,
Sivan Kartha, Robert Keohane, Donald Kraemer, David Lefkowitz, Hennie
Lötter, Eric Maskin, Michael Pendlebury, Henry S. Richardson, Lynn Rus-
sell, Axel Schafer, Nancy Schrauber, Martin Seel, Peter Singer, Angela M.
Smith, Robert H. Socolow, Richard C.J. Somerville, Steve Vanderheiden,
Paul Voice, and Michael Walzer. I am especially grateful to the extensive
comments on the entire manuscript from Henry Shue, Mary Tjiattas, and an
anonymous reviewer for Cambridge University Press.

I have also profited immensely from sharing my ideas with audience members at class presentations, conferences, and talks given at various institutions.
In the world of academic philosophy we often learn from strangers. I would
like to express my thanks to the organizers and audiences of such presentations at the American Philosophical Association Central Division meeting, Arizona State University, Bennington College, Columbia University,
the Forschungskolleg Humanwissenschaften, the Freie Universität Berlin,
Georgetown University, the Institute for Advanced Study, the International
Association for the Philosophy of Law and Social Philosophy, Johann Wolfgang Goethe-Universität Frankfurt am Main, McGill University, the North
American Society for Social Philosophy, North Carolina State University,
Oxford University, Rhodes University, San Diego City College, Technische
Universität Darmstadt, the University of Colorado, Universität Kassel, the Uni-
versity of Pennsylvania, the University of Richmond, and Washington and Lee
University.

I am thankful to my students at SDSU, both undergraduate and graduate, who discussed some of these ideas with me in courses on climate change and morality, environmental ethics, global justice, and the value of biodiversity. Having the opportunity to discuss some of this material with you has helped my research tremendously. And I am very grateful to Daniel Callies for his invaluable research assistance, copyediting help, and indexing.

Chapter 1 grew out of my "A Normative Account of Dangerous Climate Change," *Climatic Change* 108 (2011): 57–72. Chapter 2 builds on my "A Right to Sustainable Development," *The Monist* 94 (2011): 433–452. The Afterword was first published in *Dissent* online, October 31, 2012. I 'am grateful to the publishers for permission to use the earlier published material.

Finally, my deepest gratitude goes to my family, Bonnie and Marino Friedmann, who are always there for me and who keep me going.

Introduction

Our world is beset with several pressing problems, including war, intolerance, poverty, and climate change. The theme of this book takes up the last two items on this bleak list. It should be scandalous that nearly half the world's population lives in desperate poverty, especially while many lavish in such plenty. And the fact that despite the formation of the 1992 United Nations Framework Convention on Climate Change (hereafter "UNFCCC" for the institution and "the Convention" for the treaty), globally we continue to emit more CO_2 each year, even as we become more aware of the harmful effects of doing so, is evidence of an enormous collective failure. This book takes the problem of global poverty to be central to climate change policy. In so doing, it rejects the approach of many attempts that address the problem of climate change in isolation. Some even make a virtue out of doing so. I take that to be a grave moral mistake.

Some analysts of climate change policy contend that an international climate change agreement should be oriented only around the values of efficacy and efficiency. Claims of equity, fairness, and poverty eradication are sometimes rejected as redistributive and as poison to the process of reaching a climate change agreement, because redistribution would render an agreement counter to the interests of highly developed states.[1] Two considerations tell against this view. First, from the beginning of the international discussions on climate change it has been clear that climate change mitigation was one important aim, but not the only one. That aim has always been accompanied by a concern that poverty-eradicating human development be continued – that a climate change regime not establish hurdles to this developmental aim. International climate negotiations are not simply about climate change;

---

[1] This position is defended by Eric A. Posner and David Weisbach in *Climate Change Justice* (Princeton: Princeton University Press, 2010), pp. 79–88.

they are also about a fair framework for energy consumption, which is necessary for states seeking to eradicate poverty within their borders. This point is too often missed, and not only by writers who cannot see beyond the value of efficiency. Even writers who take the negotiations to be concerned with multiple problems can be blind to the importance that regulating access to energy for poverty-eradication purposes plays in the negotiations.[2] Any proposal regarding mitigation and adaptation that does not take into account the broader objective of poverty eradication will be insensitive to a fundamentally important feature of the context of climate change negotiations, namely a fair framework for energy consumption. Second, it is false that an international regime that assigns responsibility to states simply with the goal of achieving efficiency would be especially likely to generate broad agreement. If such a regime did not protect the claims of states to pursue human development, it would not find adequate support among least developed and developing states.[3]

The Convention is the first comprehensive international attempt to pull together concerned states for the purpose of avoiding dangerous climate change. The Kyoto Protocol was written, and annual Conferences of the Parties (COPs) occur, under the auspices of the UNFCCC. The COPs are large and unwieldy affairs. And because the UNFCCC has been ineffective in producing a comprehensive climate change agreement, it has come under criticism as the wrong place to expect progress on climate change.[4] Although it is important to consider these criticisms carefully, critics of the UNFCCC often overlook one of its important features. The language of the treaty is a rich source of norms and principles for guiding future deliberations and action to mitigate and adapt to climate change. This book takes these norms seriously because they provide guidance for international deliberations

---

[2] Robert Keohane and David Victor identify four problems that climate change negotiations seek to address, and access to energy for purposes of eradicating poverty is not on the list. See Robert O. Keohane and David G. Victor, "The Regime Complex for Climate Change," *Perspective on Politics* 9 (2011): 13. For an example of a proposal based not on efficiency but on a narrow conception of justice in the distribution of emissions without responding to the importance of energy for human development, see Lukas H. Meyer and Dominic Roser, "Distributive Justice and Climate Change. The Allocation of Emission Rights," *Analyse & Kritik* 2 (2006): 223–249. Approaches like this are criticized in Simon Caney, "Just Emissions," *Philosophy and Public Affairs* 40 (2012): 255–300.

[3] This point is a central theme of, and well defended in, J. Timmons Roberts and Bradley C. Parks, A Climate of Injustice: Global Inequality, North-South Politics, and Climate Policy (Cambridge, MA: The MIT Press, 2007).

[4] See David G. Victor, Global Warming Gridlock: Creating more effective strategies for protecting the planet (Cambridge: Cambridge University Press, 2011), Chapter 7.

by taking some discussions off the table and by directing the ambition of proposals made in international negotiations. Without these norms parties would have either to revisit the question of the aims of the negotiations each time they met or to tolerate discussions and proposals based on fundamental differences of value.[5] This would surely handicap and prolong the negotiating process. Whatever the weaknesses of the UNFCCC project – and there are many – the norms and principles contained in the document are important for facilitating appropriate agreement, and this book offers interpretations and defenses of several of them.

This is a book about climate change policy written by a political and moral philosopher. It assumes that the development of policy should be based, among other things, on what is important to promote and protect. Considerations of what we should promote and protect take us straight to a discussion of values. And the study of values is the bread and butter of political and moral philosophy. Climate change and climate change policy raise several important questions of value. What sort of climate change policy do we owe the poor of the world who are particularly vulnerable to climate change? Why should our generation take on the burden of mitigating climate change that is caused, in no small part, by emissions from people now dead? What value is lost when natural species go extinct, as they may well do en masse as a result of climate change? Understanding both the context in which these questions arise and what might count as good answers to them requires also understanding some of what natural scientists and economists have to say about climate change. I have no special expertise in these fields, but I have sought to read and understand the literature, not to pass judgment on debates where I have no expertise but to understand questions (and their possible answers) such as those that I just mentioned.[6] In discussions of values, moral and political philosophers can play a crucial role. With some luck, our education and experience provide us with intellectual tools that can help clarify what is at stake in discussions of values, where arguments have gone awry, and ultimately what we should do.

Professional philosophers, like many other academics, tend to write mostly for their peers in the profession. It is by means of reading and thinking hard about each other's papers that we come to understand better the questions

---

[5] On "sticky" principles, see Roberts and Park, *A Climate of Injustice*, pp. 222–223. They draw on Robert Keohane's argument about "agreement-facilitating effects of the information provided" by principles, rules, and regimes. See Robert Keohane, *After Hegemony: Cooperation and discord in the World Political Economy* (Princeton: Princeton University Press, 1984), p. 102.

[6] Only the Report of the First Working Group of the Intergovernmental Panel on Climate Change's *Fifth Assessment Report* was published at the time of this writing; so I relied on that and the *Fourth Assessment Report*.

and possible answers that animate our discipline. We hope that we are getting closer to the truth of an issue and we believe that the discipline advances by means of this activity. Often these discussions get technical very quickly, and typically they require an understanding of the dialectical context that must be built up by a great deal of background reading.

Climate change policy, however, will not be made by philosophers. It will be made by diplomats and lawmakers and in response, at least in part, to their understanding of the issues, the advice that they receive from policy analysts, and the political pressure they come under from their citizenry. One of the convictions that motivated me in the writing of this book is that moral and political philosophers have important tools to help in making sense of what is at stake in climate change policy. But if that is the case, and if we hope to advance understanding and improve policy in this regard, we had better try sometimes to present our views in a way that is understandable to people outside of the professional discipline, who might care to listen to what we have to say. This book is such an attempt. It takes up climate change not first and foremost as a philosophical problem of interest only to advanced students of the discipline, but as a public problem on which philosophy can shed some light. My hope is that the book will be of interest to non-philosophers, and especially to people who come to it because of an interest in climate change policy. This includes researchers and students of climate policy but also many others who seek a sharper understanding of the values at stake in climate change and climate change policy. But I hope that the book will also be of interest to philosophers and students of philosophy, because as more philosophers have taken up the topic of climate change, important debates are beginning within the discipline.[7]

The discipline of philosophy has various names for the kind of enterprise undertaken in this book. It is sometimes called *applied ethics* or perhaps *applied political philosophy*. This, I think, is not a very good name for what I have tried to do, given that, for the most part, I am not interested in the problems of applying principles that have been justified elsewhere at some higher level of abstraction. In one of his many books, the legal philosopher Ronald Dworkin described what he was doing as philosophy from the inside out.

---

[7] Important recent books by philosophers include Stephen M. Gardiner, Simon Caney, Dale Jamieson, and Henry Shue, *Climate Ethics: Essential Readings* (Oxford: Oxford University Press, 2010); Stephen M. Gardiner, *A Perfect Moral Storm: The Ethical Tragedy of Climate Change* (Oxford: Oxford University Press, 2011); Dale Jamieson, *Reason in a Dark Time: Why the Struggle Against Climate Change Failed and What It Means for Our Future* (Oxford: Oxford University Press, 2014); and Henry Shue, *Climate Justice: Vulnerability and Protection* (Oxford: Oxford University Press, 2014).

Rather than applying independently justified principles, he sought to advance philosophical understanding by looking at the problems and the practice first.[8] Better than *applied ethics* is, I think, *practical ethics* – or, more broadly, *practical philosophy* because this drops the connotation of applying a principle justified elsewhere. But whereas the first seems too narrow, because ethics is often construed to focus only on individual action, the latter is too broad, as practical philosophy can include all of moral philosophy.

The name that I believe fits best draws from the tradition of pragmatism in philosophy, namely *public philosophy*. This name conveys, I think, the idea that there is an attempt to talk about something of profound public importance, and to do so to an audience that is broader than only academic philosophers. If this book is a piece of public philosophy, readers will have to decide whether it successfully manages to talk about something that it is important to the public and to an audience that includes nonprofessional philosophers, in addition to addressing advanced students of the discipline.

The goal of climate change mitigation is widely regarded as the avoidance of dangerous climate change. But the efforts of natural and social scientists to provide an account of dangerous climate change have fallen wide of the mark. Chapter 1 takes up this topic. I use what I call *the personal analogy* to show that dangerous action is not a matter of risks alone, but a matter of whether conduct is too risky in light of one's values. Danger is then necessarily a normative concept, which picks out what is too risky and therefore ought to be avoided. In the context of climate change, judging what we have reason to avoid requires paying attention to three categories of reasons: the reasons that people in the future will have that we mitigate; the reasons the people presently have to consume energy to fuel poverty eradicating human development; and the reasons that people in the future will also have that we consume energy for human development. These three categories of reasons are important in subsequent discussions in the book. I argue that the relevant norm for identifying climate change is moral, and that we can adjudicate between the three kinds of reasons just mentioned by considering what respect for human dignity requires. This is the basis for my defense of what I call *the antipoverty principle*, which directs our attention to what we should avoid:

> Policies and institutions should not impose any of costs of climate change or climate change policy (such as mitigation and adaptation) on the global poor, of the present or future generations, when those costs make the prospects

---

[8] Ronald Dworkin, Life's Dominion: An Argument about Abortion, Euthanasia, and Individual Freedom (New York: Vintage Books, 1994), p. 29.

for poverty eradication worse than they would be absent them, if there are alternative policies that would prevent the poor from assuming those costs.

The danger that climate change poses to humans surely does not exhaust its dangers. Climate change could well produce mass extinctions of species, with estimates ranging from 30 percent to 70 percent of existing species. This would be a terrible loss that we should want to avoid. But the reasons for this are not the same as the reasons to avoid the prolongation of poverty. It is implausible that plant and animal species should be characterized as possessing dignity. Indeed, when one thinks about the kinds of things species are, namely closed gene pools, it might seem puzzling why they are even valuable. Chapter 2 argues that species possess presumptively high economic value. But that is not the whole of the story. This chapter introduces an idea I call *the normative gap*, which is the logical gap between recognizing something as being good or having a good and being under duty to the thing in virtue of this good. The normative gap explains why an argument for a moral duty to an organism cannot be derived from the claim that organism has a good. By enlisting the writings of the unfortunately neglected art historian and naturalist, John C. Van Dyke, the chapter also argues that organisms possess aesthetic value, the loss of which is something we have reason to avoid. Species loss is the final loss of aesthetically valuable organisms comprising the species.

One striking feature about climate change is just how much of the forecasting is riddled with uncertainty. Chapter 3 distinguishes between risk and uncertainty and also distinguishes between different sources of uncertainty: epistemic and moral. I defend a precautionary approach to climate change policy based on the rule of thumb called *the minimax rule* for deliberation in specific conditions of uncertainty, and I argue that those conditions apply in the case of climate change. The minimax rule supports a precautionary approach to climate change policy. In the context of climate change, uncertainty about grave outcomes adds to the reasons for mitigation. In this chapter I also introduce what I call *the psychological fallacy*, which occurs when one uses psychological dispositions, such as being risk averse, as models for the kinds of reasons appropriate for the justification of public policy.

Because of the long residence time of CO_2 in the atmosphere and the thermal inertia of the oceans, the CO_2 we emit today will probably have effects for hundreds, if not thousands, of years. This requires us to think in longer terms than we are accustomed. Standard approaches to long-term planning in economics seek to optimize intergenerational consumption discounted according to a social discount rate. Such a discount rate is also used when calculating the future costs of climate change. Disagreement exists

among economists about how costly climate change is, but much of this disagreement is not about what the outcomes will be. Rather, it is about how to
value them. It is, in other words, a moral disagreement. In Chapter 4 I argue
that the standard economic approach to valuing the future of costs of climate
change is highly dubious because of some of the factors used in the social
discount rate. When thinking about what we should do on behalf of future
generations to mitigate climate change, it is much more reasonable to employ
a precautionary approach, which supplements our reasons for mitigation. This
chapter also introduces the idea I call *the morally constrained CO_2 emissions
budget*. To arrest global warming, we must transition to a no-carbon economy.
For any particular temperature increase there is budget for cumulative CO_2
emissions. This idea is important in subsequent chapters.

States should not be equally burdened with the task of mitigating climate
change. The UNFCCC claims that states have the right to sustainable development. In Chapter 5 I offer my understanding of that right as a claim the
least developed and some developing states may make on highly developed
states, such that the former are able to pursue poverty-eradicating human
development without energy price increases attributable to climate change
mitigation. The defense of this right is made on the basis of two claims: the
prior commitment of parties that have ratified the Convention, and the idea
of fairness in an international framework for access to energy oriented toward
mitigating climate change.

Theories of justice give accounts of who is owed what; they are accounts
of the moral creditors. Theories of responsibility are accounts of the moral
debtors. In Chapter 6 I explain the responsibility of the present generation to
mitigate climate change as a duty not to harm the next generation by using
more than our share of the morally constrained CO_2 emissions budget. Moreover, I defend an intragenerational distribution of the responsibility to mitigate
based on a conception of responsibility consistent with the right to sustainable
development. I argue that the adequacy of a conception of responsibility
depends on the purpose to which it is put. I defend a conception of responsibility for the framework of a mitigation treaty that I call *social responsibility*.
Finally I argue for the establishment of an ability-to-pay account of social
responsibility for a climate change treaty.

Although it is a depressing thought to entertain, it nonetheless is true that
thus far we have utterly failed to mitigate climate change. Our failures make
it increasingly less likely that warming will be limited in accordance with
internationally affirmed warming limits. In light of this, it seems to make sense
to consider some alternatives. In Chapter 7 I consider the merits of three such
alternatives: increased planning for and investment in adaptation, tests into

the assisted migration of plant and animal species, and research into geo-engineering. I argue that these cannot plausibly be understood as alternatives to mitigation, but that each is an important supplement. In light of this, I return the discussion of mitigation policy and set out conditions for a morally satisfactory international mitigation policy. I argue that the version of the policy known as Pledge and Review that is incorporated into the Copenhagen Accord fails to satisfy these conditions. This does not rule out a more satisfactory version of Pledge and Review. But a general problem with such approaches seems to be a lack of urgency.

While I was writing this book, Frankenstorm Sandy ravished the Caribbean and the eastern seaboard of the United States. The Afterword contains a very brief piece on Sandy originally published in the online version of the magazine *Dissent*. Sandy may have been an important turning point in U.S. opinion regarding climate change. Then, as I was finishing this book, typhoon Haiyan – another devastating Frankenstorm – ravaged the Philippines. Because these storms may well be the new normal, I decided to include the piece as part of the book.

Finally, this book contains four appendixes. The first three are aimed primarily at advanced students of philosophy and my colleagues in the discipline. In these appendixes, I take up themes in normative ethics and justification that will mainly interest them. And owing to the more technical nature of these chapters, they are not likely to be of much interest to people who are interested in climate change policy first and philosophy only second. I have placed these discussions in the appendixes so that they can be easily skipped by readers who are less interested in philosophical debates. For those who are interested in the material, the themes of Appendixes A and B follow directly on the discussions of Chapter 1. So, they are best read between Chapters 1 and 2. The short discussion in Appendix C is best read after Chapter 5. The fourth appendix contains the Declaration on Climate Justice, which is endorsed by members of the High Level Advisory Committee to the Climate Justice Dialogue, an initiative of the Mary Robinson Foundation – Climate Justice and the World Resources Institute. The Declaration aims to mobilize political will and creative thinking to shape an ambitious and just international climate agreement in 2015. This is an important political document that is consistent with the positions that I defend in this book and I offer my support for this effort by including the Declaration within these pages.

1

Danger, Poverty, and Human Dignity

"[I]f a clod be washed away by the sea, Europe is the less."
— John Donne

It is widely appreciated that climate change is a moral problem, but perhaps not so well understood is that even identifying climate change as dangerous necessarily involves a moral judgment. This is surprising because it means that moral judgments enter into climate change policy discussions very early on, already at the point of identifying when, or what kind of, climate change is dangerous.

The stated central objective of the United Nations Framework Convention on Climate Change "is to achieve, in accordance with the relevant provisions of the Convention, stabilization of greenhouse gas concentrations in the atmosphere at a level that would prevent dangerous anthropogenic interference with the climate system."[1] If we suppose that this is the principal objective of international climate change negotiations and policy formation, we are left with some obvious questions: What is dangerous anthropogenic interference with the climate system? How is it to be identified so that policy can be crafted to avoid it?

The mean surface-air temperature of the Earth is already around 0.74°C (1.33°F) warmer than during preindustrial times. The Earth is absorbing more of the Sun's radiative energy caused by the greenhouse effect; meanwhile, the thermal inertia of the oceans causes them to warm more slowly than the mean atmospheric temperature just above land. The effect of this is that the Earth is already in for more warming even if we do not add any more CO_2 to the atmosphere, just as the water in the kettle on the stove warms more

---

[1] United Nations Framework Convention on Climate Change, 1992. http://unfccc.int/essential_background/convention/background/items/2853.php (accessed October 4, 2012).

slowly than the air immediately around the flame. We are committed to more warming, probably around 1° Celsius (1.8°F) more. Avoiding climate change altogether is not a realistic policy option. If international policy is to be guided by the Convention's central objective, a reasonable distinction between climate change and dangerous climate change is needed.

We need an identificatory account of dangerous climate change; an account that serves to identify dangerous climate change. An obvious thought is that the important work of climate scientists can provide us with such an account. Climate scientists desiring intelligently guided international policy have sought to assess the dangers of climate change, most often by discussing the risks and uncertainties associated with different warming scenarios. The risks of damage accumulate and the uncertainties loom larger as warming increases. Appreciating this is very important for intelligent policy, but it is not enough. Judging something to be dangerous relies on there being a reason to avoid it, as the language of the Convention suggests. That's not necessarily true of the judgment that something is risky.

In this chapter I defend an account of the concept of danger as too risky in light of the available alternatives, and I argue that the judgment that an action or policy is too risky involves more than an empirical estimation of the risks involved. The judgment rests necessarily on considerations of value. This is a claim that I support by developing what I call *the personal analogy*. The conception of danger that I develop relies on the value of human dignity, widely supposed to be the basis of human rights. This conception supports an approach to identifying dangerous climate change on the basis of reasons that we can all share,[2] and leads to the idea that whether poverty eradication is delayed by either climate change or climate change policy is fundamental when identifying either as dangerous.

DANGER AND VALUES

In trying to understand the nature of dangerous climate change, it would be folly not to incorporate the findings of climate science. We need to understand what CO_2 does in the atmosphere, how long it stays there, what its likely climatic effects are, how and where it returns to the Earth, and the variety of positive feedbacks that warming produces. Many climate scientists have carefully studied these matters for decades. As they continue to learn more,

---

[2] The distinction between a concept of something and the conception of it derives from John Rawls. See his *A Theory of Justice*, rev. ed. (Cambridge, MA: Harvard University Press, 1999), p. 9.

we'd better listen carefully. If policy makers don't pay attention to what the climate scientists are saying, then the policy makers certainly won't understand the risks. Not paying attention to scientists in this case would be like not listening to a doctor's assessment prior to arduous mountain climbing. If you want to know whether mountain climbing is especially dangerous for you, given your medical condition, it makes good sense to listen to what your doctor says about the risks you would assume were you to climb.

We might assume that a dangerous option is simply a risky one. But for present purposes, that will not do. If dangerous climate change is simply risky climate change, then the central objective of the Convention is impractical because the avoidance of risk altogether is not possible in climate change policy. If we take the Convention as relevant to constraining climate change, we do well to provide a construction of the central objective that is practical; one that provides the right connection to the attitude of avoidance. In this section I suggest that it makes better sense to understand the concept of dangerous climate change provisionally as climate change that is not simply risky, but *too* risky in light of the available alternatives. Doing so has two virtues. First, it preserves the practical value of the central objective of the Convention. Second, taking "dangerous" as too risky establishes the appropriate conceptual connection between a judgment of danger and an attitude of avoidance; it makes it clear that dangerous climate change is to be avoided.

Returning to the mountain climbing example, your doctor's assessment will take you only so far. The judgment that mountain climbing would be too risky is not ultimately determined by the medical facts and projections alone. It is normative. Normative judgments employ values that direct our attitudes and actions. The judgment that mountain climbing would be too risky is not merely a report about your condition. Your doctor can offer expert knowledge on your state of health and can tell you what might happen if you decide to climb. But neither the doctor's report nor prediction ultimately decides the question of whether climbing is something you should pursue or avoid. The heed paid to respect for patient autonomy in bioethics and good clinical practice is evidence of the broad acceptance of this view.

To the extent that whether you should climb is decided by your interests in, say, adventure and good health, the normative judgment about whether or not climbing is too risky – or dangerous – is a prudential one. You consider the expert opinion regarding the risks, you consider what you have reason to value, and you determine whether the risks are worth it to you. Discussions of the nature, content, and justification of normative judgments are the stock and trade of philosophy – and when the values are moral, moral philosophy in particular.

Climate scientists have approached the assessment of danger in two main ways. The dominant one involves a projection of the risks associated with climate change. In an effort to bring science to bear on the assessment of dangerous climate change, former UK Prime Minister Tony Blair hosted a major international conference on climate change in 2005 at the Hadley Centre in Exeter. Prime Minister Blair set before the attendees of the conference the task of establishing the threshold of danger: "What level of greenhouse gases in the atmosphere is self-evidently too much?"[3] The record of the conference contains papers that focus on the effects of warming on vulnerable areas of the Earth's climate system and geography, and the impact of change in these areas on humans and ecosystems. This procedure for assessing danger involves laying out the various possible bad and catastrophic effects, and discussing the probabilities that can be attached to them. This is sometimes referred to as "risk analysis." The reports of conferences such as these are exceedingly valuable for an understanding of the risks of climate change.[4]

A second approach to the assessment of danger has drawn on the findings of the social and behavioral sciences. It involves analyzing people's reactions to the possibilities of bad and catastrophic effects. The idea is to understand the psychological and cultural factors that go into a person's assessment of danger. These natural science and sociobehavioral science approaches are sometimes distinguished as external and internal approaches to danger assessment.[5]

The mountain climbing example suggests that an identificatory account of dangerous climate change – where "dangerous" is understood as too risky – is inadequate unless it can provide reasons that the climate change should be avoided. The problem for the external approach is that highlighting the bad effects and their probabilities – although an essential exercise for a rational assessment of risk – does not necessarily establish that these effects should, all things considered, be avoided. Two people can be presented with the same set of facts and probabilities about a hazardous line of employment – high-rise construction, for example – and one has reason to accept the risks, given the pay rewards, the available employment alternatives, and the need

3 Hans Joachim Schellnhuber, et al. (eds.), *Avoiding Dangerous Climate Change* (Cambridge: Cambridge University Press, 2006), vii.
4 See Stephen H. Schneider, "What is 'Dangerous' Climate Change," *Nature* 411 (2001): 17–19. See also Stephen H. Schneider and Janica Lane, "An Overview of 'Dangerous' Climate Change," in Schellnhuber, *Avoiding Dangerous Climate Change*, 7–24. See note 7. See also Brian C. O'Neill and Michael Oppenheimer, "Dangerous Climate Impacts and the Kyoto Protocol," *Science* 296 (2002): 1971–1972.
5 Cf. Suraje Dessai, et al., "Defining and Experiencing Dangerous Climate Change," *Climate Change* 64 (2004): 11–25.

for income whereas the other does not. The person accepting the job has not *necessarily* failed to appreciate the facts and risks. The question of whether the person accepting the job *should* accept it simply is not fully answered by the assessment of the facts and risks. It depends on the reasons the individual has in light of the value of the various alternatives. High-rise construction may be perfectly rational if there are bills to pay and few employment alternatives.

The problem for the internal approach is that although coming to understand which psychological mechanisms or cultural norms influenced the two people might help to predict their decisions in the future, it does not tell us whether they *should* accept or reject the job. Prediction and normative evaluation are two distinct enterprises. A normative assessment of which changes to the climate should be avoided is not then captured by either the natural or sociobehavioral scientific accounts of danger. Assessing whether or not climate change is dangerous is not simply a scientific matter. The conception of danger that is central to the judgment is normative.

I have been arguing that there is a distinction between what science can tell us about the world and what we have reason to do in light of what science tells us. Determining whether one should avoid or pursue a course action involves a scientific appreciation of the risks, but it also involves a normative evaluation, which science cannot provide. Is this too sharp a distinction between science and normative assessment? John Dewey writes that, "When physics, chemistry, biology, medicine, contribute to the detection of concrete human woes and the development of plans for remedying them and relieving the human estate, they become moral; they become part of the apparatus of moral inquiry."[6] According to Dewey, this humanization of science enriches not only science, but morality as well, which "loses its peculiar flavor of the didactic and pedantic; its ultra-moralistic and hortatory tone."[7] There are, of course, scientists – including climate scientists – who would be inclined to disagree. Some scientists maintain that responsible science requires policy neutrality. The Intergovernmental Panel on Climate Change (IPCC), for example, carefully avoids making policy prescriptions, preferring to maintain a neutral stance on what should be done about climate change. Perhaps this lends broader credibility to the science that the IPCC reports. But even so, it is not necessarily a rejection of Dewey's main claim, which might be interpreted as taking the measure of the importance of scientific inquiry to be the extent to which it results in socially relevant knowledge. Scientists themselves need not

---

[6] John Dewey, *Reconstruction in Philosophy*, enlarged ed. (Boston: Beacon, 1957), 173.
[7] *Ibid.*

be the ones pressing the social relevance of their projects. To ensure broader credibility, perhaps they sometimes should not.

Whatever the intellectual or social merits of Dewey's view, there is no reason to think that it contradicts the argument that I have been making about the insufficiency of scientific understanding to the normative assessment of dangerous climate change. I'm not at all interested in disputing the view that science should pursue morally appropriate goals. In the case of climate change, the work of climate scientists is of the utmost value to social policy. The view that I endorse about how to identify dangerous climate change involves both a scientifically informed understanding of the likely consequences of climate change and a role for moral deliberation in assessing those consequences.

THE PERSONAL ANALOGY

Examples of deliberating about mountaineering or taking risky employment rely on a personal analogy. My point in the discussion of the personal analogy is to illustrate the limits – but certainly not the irrelevance – of science in the assessment of danger. The identification of danger is not settled by any physical measurement or inference of the probability of an outcome unless one thinks that the person taking the high-rise construction job is necessarily choosing irrationally – pursuing something that necessarily should be avoided. This can be captured by saying that the normative judgment of danger is underdetermined by the empirical assessment of the probabilities of bad outcomes. Until we view the outcomes in light of the values that we hold, no judgment of danger is sensible. This discussion, of course, also shows the importance of the scientific assessment of the matter. A person would not be deliberating well if he or she ignored such an assessment. The better we understand the circumstances, the more reliable our judgments of danger will be. To judge that a course of action is dangerous involves a consideration of probabilities in light of what we have reason to value.

From the lesson of the personal analogy – that an assessment of the risks of an activity is not enough to establish whether it is too risky – we should not conclude that the assessment of danger necessarily depends on whoever is making it. If this were so, there would be little possibility for an international determination of danger. Prime Minister Blair's well-intentioned efforts would be completely misguided. People led to this view embrace a version of relativism about the standards of danger. Whatever the intellectual merits of relativism about values – and this is not the place to discuss its general merits – the idea that danger is in the eye of the beholder is not a doctrine particularly well-suited for solving the practical problem of achieving international consensus concerning what to do about climate change. More importantly, the view that judgments of danger employ values does not require us to believe that the values employed in the judgment are in any way relative to the person making the judgment. And in any case, the view that society should grant a person the discretion to choose a risky job (within limits) is well-supported by appealing to the nonrelative value of the right of the person to pursue preferred occupational and career objectives. Affirming that right does not in any way commit us to believing that every career choice a person makes is morally correct.

There are limits to the applicability of the personal analogy. The afore-mentioned reasoning might support laws that allow adults to smoke cigarettes, cigars, and pipes. But it does not support allowing them to smoke in places where the smoke can be harmful to others. (And it does not rule out taxing tobacco if there are public costs associated with smoking.) Discretion to individuals need not be extended to invading the rights of others. This is a view that is as old as liberalism in political philosophy. John Locke writes that the limits of toleration extend to practices and beliefs insofar as "they are not prejudicial to other men's rights, nor do they break the public peace of societies."[8]

CO_2 emissions are changing the Earth's climate, thereby affecting all life on the planet. In contrast to those seeking risky employment, emitters of CO_2 are not merely assuming risks for themselves, and those setting climate change policy are not merely choosing the rules to govern their own conduct. For these reasons, the personal analogy breaks down in the case of climate change; there is not the same scope for individual discretion that exists in choosing a job in high-rise construction. It would not be appropriate, for example, to permit the person choosing employment in high-rise construction the discretion to decide whether he or she may toss down tools from up high into crowded spaces below. Nor is it appropriate to let a polluter decide what level of pollution is dangerous for others.

The personal analogy helps us to see that judgments of danger are based on values. This elucidates the peculiar action-guiding character of the concept of danger. Still, analogies are tools for understanding and persuasion. And as with all tools, the use-value of the personal analogy is limited. It does not help us to understand what in particular counts as dangerous in the case of climate change. It would lead us astray if it led us to believe that the assessment of danger could be left to the particular state formulating its energy policy. The energy policy of every state – some much more than others – affects the well-being

---

[8] John Locke, "A Letter Concerning Toleration," in David Wootton ed., *John Locke, Political Writings* (Indianapolis: Hackett Publishing, 1993), 417.

of people around the globe. So, insofar as climate change is a global problem, the standard of danger cannot simply vary from state to state.

A central difference between the examples used to illustrate the personal analogy and the case of dangerous climate change is that in the former, the risks are assumed only by the individual deciding; in the latter, they are being imposed on others. Danger in the case of the personal analogy invokes only self-regarding values. In the case of climate change, it invokes other-regarding values. Different conceptions of dangers are appropriate to the two cases. The prudent person in the case of the personal analogy will decide in light of his values. But norms of prudence are inappropriate in the case of climate change. The appropriate norm there is moral. An appropriate standard of danger would have to be informed by the right sorts of concerns for those who are made vulnerable because of climate change. Otherwise, it would be morally dubious. Unlike the normative assessment of danger involved in deliberating about whether to climb mountains or to take risky jobs, the assessment of danger in the case of climate change involves deliberation not just about risks to oneself, but necessarily about risks imposed on billions of other people. A conception of danger for climate change purposes must be moral.

REASONS FOR MITIGATION AND CONSUMPTION

Direct scientific observation provides convincing evidence of a global mean temperature increase already, and there are compelling reasons to suppose such an increase has wide-ranging – but not always well-understood – climatic effects. Such warming, however, is only the beginning. The thermal inertia of the oceans will ensure that the solar radiative energy trapped by greenhouse gases will continue to warm the planet until an equilibrium point is reached, even if we were to emit no more greenhouse gases. But we do continue to emit. In fact, despite decades of publications by the IPCC of the risks of CO_2 emissions, we remain on a long-term trajectory of increased emissions. In 1990, when the IPCC issued its *First Assessment Report*, total global CO_2 emissions were 21,683.16 million metric tons (mmt). By 2009, emissions had climbed nearly 50 percent to 30,313.248 mmt.[9] These emissions are, of course, increasing atmospheric concentrations of CO_2. Owing to the long residence time of CO_2 in the atmosphere, emissions will have to decrease precipitously in order to stabilize the concentration of CO_2 in the atmosphere. As long as

---

[9] U.S. Energy Information Administration, *International Energy Statistics*. http://www.eia.gov/cfapps/ipdbproject/IEDIndex3.cfm?tid=90&pid=44&aid=8 (accessed November 22, 2012).

more CO_2 enters the atmosphere from surface sources (for example, natural decay and anthropogenic emissions) than comes out through absorption by the oceans and plant photosynthesis, the atmospheric concentration will increase even if we are reducing our emissions.

All of this adds up to the long-term accumulation of risk. The greatest risks are being assigned to persons in the future. Two factors conspire to heap the risks especially on the poor of the future. One is that large climatic impacts are expected to be experienced in tropical regions where there are a great many poor people. Another is that the poor will have fewer resources to cope with droughts, inundation by oceans, rivers flooding, tropical storms, and disease. Vulnerability to climate change can be thought of as the product of exposure to risk and capacity to adapt to adverse events. Poor people living in drought-prone regions and in large mega deltas will be made especially vulnerable by climate change. Insofar as we have good moral reasons to eradicate poverty, we have good moral reasons to avoid or at least reduce these risks.

Are there any plausible moral reasons to allow the risks – any risks – of climate change to be imposed on people, particularly the poor? Why not simply consider any climate change at all to be too risky, to be dangerous? There seem to be two main candidates for reasons that would permit some climate change. First, reducing future risks requires forgoing some benefits now; benefits associated with certain kinds of energy production. The International Energy Agency uses the term "energy poverty" to refer to the lack of access to energy – a major barrier to eradicating poverty – that a great many poor people currently confront. Around 1.4 billion people currently lack access to electricity. And 2.7 billion regularly rely on the burning of biomass, which causes indoor pollution that is particularly dangerous to children. Unless energy circumstances improve, indoor pollution is likely to become a bigger threat to human health than either HIV or tuberculosis, resulting in more than 4,000 premature deaths each day by 2030.[10] Perhaps the benefits of increased fossil fuel use to persons experiencing energy poverty are sufficient to offset the risks that would be heaped on people in the future by such use. Such a justification for passing along climate change-induced risks bears a similarity to the mountain climber concluding that the risks associated with climbing pale in comparison to what can be gained from the experience of the climb. There is, of course, a very important disanalogy to the mountain climbing example. In the case of climate change, the foregone benefits and risks often accrue to

---

[10] International Energy Agency, *Energy Poverty: How to Make Modern Energy Access Universal*, 2010. http://www.unido.org/fileadmin/user_media/Services/Energy_and_Climate_Change/ Renewable_Energy/Publications/weo2010_poverty.pdf (accessed November 22, 2012).

different people, including people in different generations. The values at stake are moral rather than prudential.

There would be a second reason to permit some climate change if the risks of significant delays in human development caused by the increased energy costs of a mitigation policy were sufficiently high. If mitigating climate change involves slowing the process by which billions of people will climb out of poverty, perhaps people in the future are better off with less mitigation. Overcoming energy poverty today will improve the lives of people in the future. Putting ourselves in their shoes, we judge there to be good reasons not to want to retard or delay the long-term process of human development by increasing energy costs, a process that will yield capital investments from which they will benefit.[11]

Identifying an appropriate conception of danger for climate change policy purposes involves assessing at least three categories of reasons:

1. The reasons that future persons would have to avoid the risks of climate perturbations.
2. The reasons that present people have to consume energy, especially to eradicate poverty.
3. The reasons that future people would have that poverty eradicating human development commence, continue, or accelerate.

Neglecting the important disanalogy mentioned earlier, the first category of reasons is analogous to the reasons a person has for not embarking on a risky course of action. Within this category are reasons for mitigation in the case of climate change. In contrast, within the second and third categories are reasons analogous to those one has to assume the risks because the alternatives appear even more risky. Reasons of these two categories in the present context are reasons for energy consumption.

As I argue in this chapter, the personal analogy is useful, but ultimately inadequate for a conception of danger in the case of climate change. Consideration of the differences between danger in the case of climate change and the personal analogy draws attention to two morally important features of the identification of dangerous climate change. The first is that it requires an intergenerational assessment; the aforementioned categories of reasons concern the interests of persons now living (or living in the near future), as well as more-distant future persons. The second feature is that the relevant risks fall

---

[11] This second kind of reason is pressed by Bjorn Lomborg in *Cool It: The Skeptical Environmentalist's Guide to Global Warming* (New York: Alfred A. Knopf, 2008).

unequally on persons within a generation. The global poor are most vulnerable both to climate perturbations and to increased energy prices.

The IPCC's *Fourth Assessment Report* (AR4) provides a brief summary of the future risks of climate change at various temperature levels.[12] It predicts with high confidence (meaning it believes that the chance of the prediction being correct is about 8 in 10) the following outcomes for particular warming levels: A 1°C temperature increase would put up to 30 percent of existing animal and plant life at increased risk of extinction and would increase coral bleaching; a 2°C increase would result in increased damage from floods and storms and in most coral being bleached; a 3°C increase would expose hundreds of millions of people to increased water stress; a 3.25°C increase would bring about increased morbidity and mortality from heat waves, floods and droughts; and a 3.75°C would cause a 30 percent loss of coastlines and the suffering of millions of people because of flooding.[13] But recent assessments of the AR4's predictions in light of observed change suggest that the risks of more significant damage may be greater.[14]

There can be little doubt that people in the future would have good reasons for mitigation. They will have interests in maintaining food and water security; avoiding damages caused by tropical storms, flooding, and rising seas; and species loss will be at least as important to them as to us. These risks must be considered in light of the risks of an international mitigation regime that would reduce them. The primary risk of such a regime is that poor people now or in the near future will lose access to energy because of increased prices. Poor people have very good reasons to want to increase their energy consumption. Electrification saves lives in hospitals and lights up homes, schools, and workplaces. Mass industrial production is not possible without massive energy use in production and transportation. Health, education, and good-paying jobs are all strongly dependent on energy consumption. A loss of access to energy would negatively affect human development efforts and have effects for human development that would extend into the future. An

[12] At the time of the writing of this book the *Fifth Assessment Report* has not been released in its entirety. In the material released, Working Group 1's *Physical Science Basis*, there is no corresponding table of risks for various temperature increases.

[13] Intergovernmental Panel on Climate Change, *Climate Change 2007: Synthesis Report Summary for Policy Makers*, 10. http://www.ipcc.ch/publications_and_data/ar4/syr/en/main.html (accessed November 22, 2012).

[14] Cf. Stefan Rahmstorf, et al., "Recent Climate Observations Compared to Predictions," *Science* 316 (2007): 709. See also Katherine Richardson, et al., *Synthesis Report*, proceedings from the conference "Climate Change, Global Risks, Challenges, and Decisions" at the University of Copenhagen March 10–12, 2009. http://climatecongress.ku.dk/pdf/synthesisreport (accessed March 4, 2010).

international mitigation plan that provides reasonable assurances that human development would not be negatively affected is then necessary to make a strong case that climate change is dangerous, at least with respect to reasonable human concerns. Just as in the case of high-rise construction, there must be a reasonable alternative before the course of action is deemed dangerous.

HUMAN DIGNITY AND POVERTY ERADICATION

In the case of the personal analogy, we think that when the rights of others do not come into play a person identifies danger by considering the risks of the alternatives in light of the individual's values. The person has a right to pursue risky courses of action and we allow considerable latitude (because individual goals vary) before we question the rationality of the decision or the individual's sanity. But energy and climate change policy concern the lives and well-being of persons in different generations, and persons who are relatively privileged as well as deeply impoverished. I discussed three categories of reasons that are salient when determining whether climate change is dangerous – or too risky in light of the alternatives. And I argued that the threshold for what is too risky must be justified in light of alternative policies taking into account the reasons for mitigation and for energy consumption.

The personal analogy lends credence to the idea that the appropriate attitude toward risks and sacrifices depends on what is justifiable to those that bear them. When it comes to matters of policy, the idea that the risks and sacrifices that a policy imposes must be justifiable to those that bear them is a fundamentally moral notion. One early expression of this idea, albeit with particular stress laid on rational agency, can be found in the writings of Immanuel Kant, who claims that the dignity of a rational being requires that "his maxims must be taken from the viewpoint that regards himself, as well as every other rational being, as being legislative beings."[15]

A monumental achievement of the public international law of the second half of the twentieth century is that it is fundamentally committed to the view that "All human beings are born free and equal in dignity and rights. They are endowed with reason and conscience and should act towards one another in a spirit of brotherhood."[16] Seeing human beings as possessors of inherent dignity involves respecting one another as coauthors of the rules of

---

[15] Immanuel Kant, *The Grounding to the Metaphysics of Nature*, 438, James W. Ellington trans. in Immanuel Kant, *Ethical Philosophy* (Indianapolis: Hackett Publishing, 1983), p. 43.

[16] Universal Declaration of Human Rights, Article 1. http://www.un.org/en/documents/udhr/index.shtml (accessed September 13, 2012).

our common life. In accordance with this inspiring moral vision, the risks and sacrifices of international policy must then in principal be justifiable to each person affected by the policy.[17] In light of the unavoidable influence that climate change policy can have on matters that we have very good reasons to value, such as life and basic well-being, imposing climate change-related risks on people threatens serious injustice.

The three categories of reasons suggest the possibility of two different kinds of dangerous mitigation policies. One kind of policy would be deemed too risky by possible future persons because the mitigation that it requires is insufficient; the greenhouse gas emissions it would allow would cause great suffering, loss of life, and a deepening of the scourge of poverty. Another policy would be judged dangerous by people now and in the future because it raises the price of energy and thereby slows poverty-eradicating human development that benefits both the impoverished alive now and their descendents. Coauthors of the rules of a common life would seek a principle that would guide policy away from these kinds of outcomes.

A few more things need to be said about this idea to make it sufficiently clear. Not everyone could be expected to endorse the kind of principle I mention. But we should aim for a principle that is acceptable to everyone seeking agreement that is based on the respect for human dignity.[18] Whether a principle is acceptable will depend on the costs and impositions it lays on people. The assessment of possible principles must consider whether such costs and impositions are acceptable to people, not just in light of what they want for themselves or others, but also in light of the reasons persons have who are seeking an agreement based on the value of human dignity. Assessing the principle, then, will require considering reasons that can be shared by others; and these should be reasons that abstract from the desires and concerns that are relevant only to an individual's own life. Sensitivity to matters that we all have reason to value and to the various ways in which policy can positively or negatively influence human lives is required.

---

[17] This is a rich idea that appears in many accounts of moral and political philosophy influenced by Kant. An important and very influential source is the following: "[R]espect for persons in shown by treating them in ways that they can see to be justified." John Rawls, *A Theory of Justice*, rev. ed. (Cambridge, MA: Harvard University Press, 1999), p. 513. For recent treatments, see Stephen Darwall, *The Second Person Standpoint: Morality Respect and Accountability* (Cambridge, MA: Harvard University Press, 2006); Rainer Forst, *The Right to Justification*, Jeffrey Flynn, trans. (New York: Columbia University Press, 2011); and Darrel Moellendorf, *Global Inequality Matters* (Basingstoke: Palgrave Macmillan, 2009), chp. 1.

[18] The account of these few paragraphs is strongly influenced by T. M. Scanlon in *What We Owe to Each Other* (Cambridge: Harvard University Press, 1998), chp. 5. But Scanlon does not base his account on respect for human dignity.

THE ANTIPOVERTY PRINCIPLE

Everyone has a reason to avoid involuntary poverty. So, if the costs of risk reduction must be borne, it would be unreasonable for the poor to bear them insofar as this perpetuates their poverty. The risks of an energy policy should neither increase overall poverty nor delay eradicating it. Because it is not possible for everyone to avoid the costs of climate change, a principle that assigns the costs to the nonpoor is the most reasonable. This applies equally to present as well as the future poor. These considerations suggest the following antipoverty principle:

> *Policies and institutions should not impose any costs of climate change or climate change policy (such as mitigation and adaptation) on the global poor, of the present or future generations, when those costs make the prospects for poverty eradication worse than they would be absent them, if there are alternative policies that would prevent the poor from assuming those costs.*

Policy that contravenes this principle is dangerous. The general idea is that we have good moral reasons to make not prolonging poverty a constraint on climate change policy. Any energy policy – whether business-as-usual or some combination of mitigation and adaptation – based on a principle that assigns costs in a way that prolongs global poverty is unreasonable.[19]

Climate change will have significant costs for future generations and the poor among them are particularly vulnerable. Our generation is not in a position to ensure that no future costs will be paid by the global poor, but we can do two things that provide some protection for them. We can assume some of the total package of costs through mitigation policies and investments in infrastructure for adaptation. And we can ensure that the poor now do not suffer more as a result of assuming these costs. Not increasing the misery of the impoverished now is, of course, morally important for its own sake, but it also ensures that the starting point for the descendants of the present generation is not as bad as it might otherwise be.

Now, it is possible that owing to the limitations of knowledge, inabilities to coordinate policy, and incapacity to pull the required policy levers, some people will be thrown into poverty or be more deeply impoverished no matter whether international policy permits climate change or establishes an international climate change mitigation regime. The antipoverty principle identifies

---

[19] This is similar to a point made by Henry Shue about whether poor countries with leverage should cooperate. See his "The Unavoidability of Justice," in Andrew Hurrell and Benedict Kingsbury, eds., *The International Politics of the Environment: Actors, Interests, and Institutions* (Oxford: Oxford University Press, 1992), pp. 381–384.

policies as dangerous if they impose avoidable poverty-prolonging costs on the poor. It is possible that not all such costs can be avoided. If one climate change-policy option prolongs or deepens poverty by a measure of X, and the other by a measure of $X - n$, and if these are the only two options, then it is only the difference, namely n, that is avoidable. The policy option of prolonging or deepening poverty by a measure of X is dangerous. If all available policies prolong or deepen poverty, the antipoverty principle identifies the one that does so the least as the one that is not dangerous.

In defending the antipoverty principle, I assume that it is coherent to affirm moral duties even when the concern is about possible future persons, whose existence is contingent on the policies that we pursue. There is considerable debate in the philosophical literature about how to understand such duties. For a discussion of this in relation to the antipoverty principle, see Appendix A.

TEMPERATURE TARGETS

At this point it may be useful to summarize the account of dangerous climate change supported by the considerations I have discussed. Dangerous climate change is climate change that is too risky, as understood with respect to both the reasons for energy consumption and the reasons for mitigation. When fossil fuel consumption is likely to make the global poor of the future worse off than they would be under a mitigation plan, and the costs of the mitigation plan would not be assumed by the global poor of this generation, then it is reasonable to avoid the risks of fossil fuel consumption. This account of dangerous climate change involves more than assembling the scientific case for the risks; a feasible alternative policy course that does not render the poor worse off is required.

Internationally, there is a consensus that 2°C is the threshold for dangerous climate change. This has been advocated by internationally respected non-governmental organizations (NGOs), and it was formally accepted at the Conference of the Parties (COP) 16 in Cancun in 2010.[20] The risks of water and food insecurity, flooding, and intense tropical storms beyond 2°C are alarming. But the Alliance of Small Island States and some least-developed states have argued that the more appropriate threshold is below 1.5°C.[21] To some island

---

[20] United Nations Framework Convention on Climate Change, *Report of the Conference of the Parties of It[s] Sixteenth Session.* http://unfccc.int/resource/docs/2010/cop16/eng/07a01.pdf#page=2 (accessed November 1, 2012).

[21] See Alliance of Small Island States, "Small Islands Call For Research On Survival Threshold." http://aosis.org/small-islands-call-for-research-on-survival-threshold/ (accessed November 5, 2012).

states, sea-level rise presents an existential threat. Given the warming we are committed to because of the thermal inertia of the oceans, the costs of limiting warming to 1.5°C might be very high, possibly resulting in a protracted global recession that would be very damaging to the global poor.[22] That also should be avoided. Any reasonable mitigation plan would need to direct its costs so as to avoid forestalling poverty eradication. The 2°C temperature limit has been accepted largely because of important scientific forecasts about the risk of serious costs and abrupt changes to the climate system if warming should go higher.

No temperature limit, however, can be accepted simply on the basis of the climate change caused risks to future generations. All limits must be assessed in light of the risks that they might pose to human development now and in the future. The identification of climate change as dangerous requires a reasonable mitigation plan without which it is unclear whether avoiding climate change would unacceptably delay poverty eradication. We cannot know at what temperature level climate change is dangerous until we have good reasons to believe that climate change is worse for the global poor than mitigation. Ultimately, a commitment to avoid dangerous climate change must also involve a commitment to a mitigation plan that avoids heaping avoidable costs on the poor. Limiting warming to below 2°C makes good moral sense only if there is also a commitment to ensure that the costs of doing so do not prolong poverty.

I turn now to a brief discussion of alternative accounts of danger. Two deserve the most attention. My suggestion is that to the extent that there is value in these approaches, it is best captured by the account that I have offered earlier.

THE HUMAN RIGHTS APPROACH

Some moral assessments of climate change lay emphasis on the threat of massive human rights violations that it poses.[23] We need only note that a United Nations Development Programme (UNDP) report forecasts an *additional* 600 million people suffering from malnutrition by 2080 to get a sense of the scale of the problem.[24] Article 22 of the Universal Declaration of Human Rights recognizes that, "Everyone, as a member of society, has the right to social

---

[22] Such a possibility is raised by Nicholas Stern in *The Global Deal: Climate Change and the Creation of a New Era of Progress and Prosperity* (New York: Public Affairs Books, 2009), p. 39.

[23] See for example Simon Caney, "Cosmopolitan Justice, Rights, and Global Climate Change," *Canadian Journal of Law and Jurisprudence* XIX (2006): 255–278.

[24] UNDP 2007–2008, 27–30.

security . . . " and Article 25 asserts that, "Everyone has the right to a standard of living adequate for the health and well-being of himself and of his family, including food."[25] It is hard for most people to imagine human rights violations on the immense scale suggested by the UNDP warnings, and we might be reticent to consider that the energy policies of various states – our states – are causing human rights violations on such a scale. But weakness of the imagination and reticence to consider our own possible culpability should not stand in the way of serious moral assessment. Indeed, some have argued that climate change should be thought of as dangerous precisely when it surpasses a threshold of human rights violations.[26]

Certain formulations of a basic right to subsistence, possibly in conjunction with other rights, would seem to line up more or less with the antipoverty principle that I previously discussed. Both human rights and the antipoverty principle have a basis in respect for human dignity. If we were to express what is dangerous about climate change in terms of the possibility of the human rights violations, it seems likely that we would arrive at judgments about danger similar to those yielded by the antipoverty principle. In fact, it might be reasonable to express the harms of avoidable poverty that make climate change dangerous according to the antipoverty principle as human rights violations. There is no fundamental contradiction between identifying dangerous climate change by means of employing the antipoverty principle and an account of the harms of climate change as human rights violations.

When it comes to identifying dangerous climate and policies to be avoided, however, invoking human rights violations alone could be inadequate because there might be no policy options that do not result in human rights violations. We would then be in need of a guiding principle that relies on something other than, or in addition to, human rights. Climate change mitigation policies may also cause suffering in ways that are similar in kind to the depravations caused by climate change itself. Mitigation policies that increase the price of energy in least-developed and developing countries might result in great hardship and even death. These depravations would be the direct result of policy and therefore, plausible candidates for human rights violations. The antipoverty principle provides a basis for identifying dangerous policies when there are no options that do not violate human rights. If there are only two policy options, and if one deepens poverty by a measure of X and the other by

---

[25] Universal Declaration of Human Rights, 1948. http://www.un.org/Overview/rights.html (accessed November 22, 2012).

[26] Rajendra Pachauri stresses the importance of rights as part of the assessment of danger. Cf. Rajendra Pachauri, "Avoiding Dangerous Climate Change," in Schellnhuber, *Avoiding Dangerous Climate Change*, 3–5. See note 2.

X – n, then the antipoverty principle identifies the former as dangerous. Further philosophical discussion of human rights and future climate change can be found in Appendix B.

AN ECONOMIC APPROACH

Another approach to assessing danger can be found in economic risk-benefit analyses. The idea is to identify climate change as dangerous when the risks of energy generated by burning fossil fuels exceed the benefits. A risk is taken to be the cost of a bad outcome multiplied by the probability of its occurrence. One virtue of such an approach is that it applies the risk-benefit assessment to both climate change and mitigation policies. The methods of risk-benefit analyses are employed in most standard economic analyses of climate change, which seek to weigh the expected utilities of alternative policies across generations in order to find the optimal one.[27] Expected utility is the value or disvalue of an outcome multiplied by its probability of occurrence. Utility, in the standard economics of climate change, is taken to be consumption, or gross domestic product (GDP), and disutility loss of consumption. The expected utility is aggregated over an infinite time horizon, with the goal of finding the optimal consumption path beginning now and extending infinitely. Both energy policy that boosts consumption and projected climate perturbations that reduce it are factored into the aggregate. Because a discount factor is applied to future utility, the approach is commonly referred to as *discounted utilitarianism*. Rather than seeking a simple equilibrium point of risks and benefits, the discounted utilitarian approach seeks to maximize over time the expected benefits and minimize the risks (measured in terms of probable consumption gains and losses) of an energy policy.

The account of identifying danger that I have presented in this chapter can be partially translated into this approach to risk assessment. My account can be seen as taking poverty eradication as having a presumptive, extraordinarily high value; so any real policy option with a significant probability of hindering poverty eradication would be too risky, or dangerous, in the sense that that the risks would outweigh the benefits. This way of expressing the position may lead to criticisms by persons working the handle of a hypothetical-scenario pump. If a billion people not in poverty would be very likely to enjoy improved

---

[27] A well-developed version of this approach can be found in William Nordhaus, A *Question of Balance: Weighing the Options on Global Warming Policies* (New Haven and London: Yale University Press, 2008). For an economic critique of seeking an optimal approach, along the lines of Nordhaus's, see Martin L. Weitzman, "A Review of *The Stern Review on Economic of Climate Change*," *Journal of Economic Literature* XLV (2007): 703–724.

social conditions, but this would require the low but not insignificant probability that poverty eradication would be postponed by one day, wouldn't it be unreasonable to deny the billion this opportunity? One way to answer this fantastical scenario relies on our ordinary, reasonable moral commitment to eradicate poverty. To the extent that we take this project seriously, we are simply committed to the "unreasonable" outcome imagined in the hypothetical scenario. And the understanding of danger simply elucidates this ordinary moral commitment to eradicate poverty.

However, there are three good reasons that we need not worry much about being unreasonable by affirming the account of danger presented here. First, we have good moral reasons for the commitment to eradicate poverty. Any person appreciating the possible misfortunes of human life knows that everyone has a reason to avoid involuntary poverty. The imposition of a climate change–risk-reduction policy based on a principle that would prolong poverty would not be acceptable. Second, there is no alternative expression of the commitment to eradicate poverty that is morally plausible. The hypothetical scenario seeks to cast doubt on the requirement not to prolong poverty by considering the hardness of a limit that rules out extending poverty by one day, or one hour, or one minute. But there is no way to state the commitment to eradicate poverty that will allow for soft limits. The principle to eradicate poverty *most of the time* not does carry much commitment and it is not even clear what it would mean – 70 percent of the time? Fifty-one percent? Third, there are no practicable policy levers that would present us with the option envisaged in the hypothetical scenario. We have no real choice to make. We would do best to stick with reasonable moral commitments about this world. The commitment not to hinder poverty eradication is a very good one given our circumstances.

The account of danger that I have presented in this chapter, however, cannot be fully translated into the economic approach. I keep my comments limited to one reason for this claim here, but I shall come back to this issue in Chapter 4. The reason has to do with the conception of what is at risk in climate change policy. Discounted utilitarianism seeks to fit all the morally relevant risks of climate change into the category of suboptimal consumption. By way of contrast, consider that the UNDP understands human development to consist of attainments in income, health, and education. Health and education attainment, in other words, are taken as goods independent of income. Both health and education might be thought of as resources, broadly speaking in that the attainments are objectively measurable, but their measure is not reducible to a measure of income. And they are not considered valuable just insofar as they boost a person's income or consumption.

The UNDP's account better captures why we have moral reason to care about human development. The example of education illustrates this. It would be wrong to think of education as merely an economic good because not only does it train people to be productive members of the economy, it is also necessary for autonomously directing one's life and becoming a good citizen. This is not to deny the usefulness of employing an economic analysis as part of the deliberative process of deciding whether to build a school or to keep one open. In a world with opportunity costs we are quite reasonably concerned about the economic costs of acting on what we value, even if its value is not exhausted by its economic costs. But only a deeply impoverished moral theory could justify the judgment that education is valuable primarily as a means to boosting consumption. Any plausible risk-benefit approach will have to turn to other matters of value.

The pursuit of optimal consumption across generations endorses continued net economic growth indefinitely. But to the extent that economic growth is valuable, it is only as a means to the improvement of people's lives and there is no reason to suppose that such improvements require perpetual growth. There are a variety of reasons to think that when achieving a high level of human development, a steady state economy without continued growth would be preferable to indefinite growth. Freedom from unnecessary toil stands out as one reason. This is the basis of John Stuart Mill's recommendation of a steady state economy:

> I confess I am not charmed with the ideal of life held out by those who think that the normal state of human beings is that of struggling to get on; that the trampling, crushing, elbowing, and treading on each other's heels, which form the existing type of social life, are the most desirable lot of human kind, or anything but the disagreeable symptoms of one of the phases of industrial progress. It may be a necessary stage in the progress of civilization ... But it is not a kind of social perfection which philanthropists to come will feel any very eager desire to assist in realizing ... [T]he best state for human nature is that in which, while no one is poor, no one desires to be richer, nor has any reason to fear being thrust back by the efforts of others to push themselves forward.[28]

More recently, several ecological economists have argued that economic growth must flatten out eventually in order to avert massive ecological

---

[28] John Stuart Mill, "Of the Stationary Sate," Book IV, Chapter VI in *Principles of Political Economy: With Some of Their Applications to Social Philosophy* (London: J. W. Parker, 1848). http://www.econlib.org/library/Mill/mlP61.html#Bk.IV,Ch.VI (accessed November 22, 2012).

destruction.[29] Success at global poverty eradication could take away the primary justification for economic growth. In the meantime, managing growth to prevent environmental catastrophe is imperative.

VALUE PLURALISM

In this chapter I have argued for a conception of dangerous climate change that is normative. Generally, we have good reason to avoid that which is dangerous. The values implicit in judging climate change as dangerous are moral. In light of what we owe other persons, we have good moral reason to avoid climate change that would prolong poverty eradication and we have reason to seek a mitigation policy that avoids this. Avoiding dangerous climate change must involve a mitigation plan that does not assign the costs of mitigation in a manner that forestalls the important moral project of eradicating poverty.

Moral values, as I have explained them here, concern how we live together in recognition of the inherent dignity of each of us. Public policy that is morally justified is acceptable to all from the point of view of the inherent dignity of each. Moral values, as I understand them, are understood from within the first-person plural perspective.[30] Presumably, there are many such moral values, certainly not only those relevant to the fair distribution of the costs and benefits of our energy policies both across countries and generations. But not all values relevant to climate change policy are accessible from the first-person plural perspective unless the first-person plural pronoun is understood to include our nonhuman natural surroundings. That seems a stretch of the use of the term "we." But no account of the values relevant to climate change policy would be complete without a discussion of the effects of climate change on the natural environment. Surely part of the reason that climate change is dangerous is that it threatens massive species extinction. I turn to that matter in Chapter 2 and return to considerations of human value in the subsequent chapters.

---

[29] Cf. Herman Daly, *Steady State Economics*, 2nd ed. (Washington, DC: Island Press, 1991).

[30] The defense of this view of morality is beyond the scope of this book, but it is defended by a variety of philosophers. See, for example, Jürgen Habermas, "Discourse Ethics: Notes on a Program of Philosophical Justification" in *Moral Consciousness and Communicative Action*; Christian Lenhardt and Sherry Weber Nicholsen trans. (Cambridge, MA: The Massachusetts Institute of Technology Press, 1991); Scanlon, *What We Owe to Each Other*; Darwall, *The Second Person Standpoint*; and Forst, *The Right to Justification*.

2

The Value of Biodiversity

"Sweet are the uses of adversity,
Which, like the toad, ugly and venomous,
Wears yet a precious jewel in his head."
— William Shakespeare

In Chapter 1, my concern was with the identification of the danger that climate change poses to humans and our societies. When the risks of climate change are imposed on people who are impoverished, but there are mitigation alternatives that would avoid that, then climate change is dangerous and should, as a matter of morality, be avoided. The basis for this account of danger is the inherent dignity of persons and the ideal of taking one another as coauthors of the rules of our common life. At the end of Chapter 1, I noted that climate change raises other considerations of value that are not well understood according to the framework of that chapter. A pluralism about values comes to the fore as we attempt to understand what is at risk in our changing climate.

Climate scientists, ecologists, oceanographers, and field biologists tell us that nonhuman life is also in peril because of climate change. This is often represented by the iconic picture of the lonesome polar bear on a diminished ice sheet surrounded by frigid seas. Where will that bear go when the polar ice has melted? However, the danger that concerns scientists is not to that individual bear, but to *Ursus maritimus*, his species, and hundreds of thousands, maybe millions, of other species that comprise terrestrial biodiversity. The enormous diversity of species – what scientists refer to as "biodiversity" – will be compromised as ecosystems approach new equilibrium points caused by changing precipitation and storm patterns, adjustments to seasonal calendars, and altered mean temperatures. Species are, and will continue to be, under new stresses. In all likelihood a great many, perhaps a majority, will not survive.

Over the very long term, owing to evolutionary pressure, new species can be expected to arise, but for millions of years life on Earth is likely to be far less rich. In terms of sheer geographic and temporal scale, *Homo sapiens* would then have left a footprint that dwarfs that of any other species. This has lead the Nobel Prize-winning chemist Paul J. Crutzen and other scientists to refer to our age as "the anthropocene" because the Earth's history is now irretrievably marked by human activity.[1]

There is something disquieting about considering the immensity of our effect and the permanent nature of the loss of species that have evolved over time horizons that we cannot even imagine. It stretches our evaluative capacities beyond everyday understandings of the consequences of our actions on individual animals or pieces of the landscape. It brings us to the edge of the territory in which the religious imagination often feels at home. Indeed, the theologian Michael S. Northcott claims that, "Adequate repair of this situation necessitates the recovery of the pre-Enlightenment Earth story; that the diverse species of the earth and the biological laws which direct life to original abundance are not the outcome of instrumental and autonomous processes, but of a beneficent divine Creation."[2] In this chapter I suggest an alternative approach. Although there are surely reasons that derive from the world's great religions to preserve species,[3] these could not plausibly generate sufficient agreement for policy in a world marked by significant religious diversity and all-too-frequent hostility. Praise for "[n]ature glorious with form, color, and motion," as Ralph Waldo Emerson intones,[4] is often sung in a religious key, but such praise is no stranger to the field biologist or to the laboratory scientist gazing through a microscope. Concern, delight, wonder, and awe at nature, beautiful and sublime, are attitudes equally available to religious and secular understandings of the world. Moreover, our climate change policy can only be improved by greater scientific understanding of the changing planet. A call to

[1] See the survey article by Elizabeth Kolbert, "Enter the Anthropocene – The Age of Man," *National Geographic*, March 2011. http://ngm.nationalgeographic.com/2011/03/age-of-man/kolbert-text (accessed December 10, 2012).

[2] Michael S. Northcott, *A Moral Climate: The Ethics of Global Warming* (New York: Orbis Books, 2009), p. 166.

[3] For example: "But ask the beasts, and they will teach you; the birds of the sky, and they will tell you; or speak to the earth and it will teach you; the fish of the sea, they will inform you. Who among all these does not know the hand of the Eternal has done this?" (Job 12:7–9); "Consider the lilies how they grow: they toil not, they spin not; and yet I say unto you, that Solomon in all his glory was not arrayed like one of these" (Luke 12:27); "Greater indeed than the creation of man is the creation of the heavens and the earth" (Quran, Surah 40:57).

[4] Ralph Waldo Emerson, *Nature*, sec. 5 "Discipline." http://oregonstate.edu/instruct/phl302/texts/emerson/nature-emerson-a.html#Introduction (accessed December 10, 2012).

retreat from Enlightenment science is particularly poor advice as we struggle to adapt to climate change.

The 1992 Convention on Biodiversity asserts "the intrinsic value of biological diversity and of the ecological, genetic, social, economic, scientific, educational, cultural, recreational and aesthetic values of biological diversity and its components."[5] This shotgun approach to valuing biodiversity is evidence of the variety of our concerns, even if it contains a confusion of categories, a blurring of the distinction between general categories (such as intrinsic and non-intrinsic) and various more specific kinds of values (which may be either intrinsic or non-intrinsic). The United Nations Framework Convention on Climate Change (UNFCCC) does not explicitly express the value of biodiversity, but it does have as an objective the stabilizing of greenhouse gases in the atmosphere within a time frame that would "allow ecosystems to adapt naturally to climate change."[6] It is not clear what is *natural* about adaptation in the context of climate change, but insofar as it is the species within ecosystems that are at risk because of climate change, it seems appropriate to read this objective as seeking to prevent (at least) mass extinction. The legal recognition of the value of species is not a peculiarity of international law. The U.S. Endangered Species Act also affirms a plurality of values, holding that "these species of fish, wildlife, and plants are of esthetic, ecological, educational, historical, recreational, and scientific value to the Nation and its people."[7]

There is scientific and philosophical controversy about the nature of species.[8] In this chapter I take the standard approach of many biologists and consider species to be simply closed-gene pools because this is consistent with common measures of biodiversity. The biological rule of thumb is that reproduction does not occur across species, or when it does, the offspring is not fertile. Species are then relatively stable units of life, temporary fixed points in a larger evolutionary process in which the random mutation of genes, adaptation, and struggle for survival propel biological change. Looked at this way, it is not obvious that concern about species should be different from genetic material generally or the individual organisms that constitute a species. But, as the previous legal passages evince, we take species to be special.

5 http://www.cbd.int/doc/legal/cbd-en.pdf (accessed March 7, 2012).

6 United Nations Framework Convention on Climate Change, Article 2. http://unfccc.int/ files/essential_background/background_publications_htmlpdf/application/pdf/conveng.pdf (accessed March 7, 2012).

7 Endangered Species Act of 1973, Sec. 2, para. 3. http://epw.senate.gov/esa73.pdf (accessed March 8, 2012).

8 For a survey of the controversy, see the entry "Species" in the *Stanford Encyclopedia of Philosophy*. http://plato.stanford.edu/entries/species/ (accessed December 7, 2012).

The judgments about species and the biodiversity expressed in the afore-mentioned legal instruments take species as valuable in a plurality of ways. In this chapter I offer an interpretation of the concern about biodiversity loss that makes sense of the plurality of valuations appropriate to species. It seems to me that there are several objectives that such an interpretation should fulfill. First, and perhaps obviously, it should take biodiversity as valuable. Although this is perhaps an obvious goal of such an account, meeting it turns out to be a significant challenge because the importance of species, as collections of plants and animals, is not well captured by approaches that focus on the moral interests and rights of individual organisms. Second, because the attitudes of concern, delight, wonder, and awe at nature are widely held, an account of the value of biodiversity should be consistent with these attitudes and should be available for political purposes in a way that is independent of whether or not the attitudes derive ultimately from a religious worldview. With the goal of international policy agreement in mind, although religious accounts are important for the mutual understanding of those who share a faith, they are ill suited as bases of agreement among people in conditions of religious pluralism. Third, assuming that the legal documents cited previously in fact do express a widely held plurality of valuations regarding biodiversity, an account of its value should try to make sense of that pluralism. And fourth, it should, to whatever extent possible, explain, and not merely express, the value of species and biodiversity.

I begin with a brief discussion summarizing the key aspects of the current natural scientific understanding of biodiversity and the threat that climate change poses to it. After that, two sections discuss the economic value of species and the limitations of this approach. Because it seems implausible that species could be merely of economic value, in the subsequent two sections I discuss accounts based on respect for nature and the land ethic. Two sections follow in which the aesthetic value of biodiversity is presented and defended. The final section returns to the theme of value pluralism, this time in the context of trade-offs between preserving biodiversity and respect for the dignity of persons.

BIODIVERSITY AND SPECIES EXTINCTION

There is both a tremendous resilience and a frightening contingency to life on our planet. Resilience is a feature of life itself, but the contingency is a feature of any particular species, including our own. The Earth has suffered five mass extinctions over the past 500 million years: the Ordovician 440 million years ago, the Denovian 365 million years ago, the Permian 245 million years ago, the Triassic 210 million years ago, and the Cretaceous 65 million years ago.[9] Paleo-scientists believe that the greatest of these was the Permian in which 77–96 percent of all marine animal life was wiped out.[10] Each of these extinctions not only threatened life, but massively denuded biodiversity. In principle, there are multiple possible measures of biodiversity. It could, for example, be measured in terms of genetic material. But the standard scientific measure of biodiversity is the number of living species. This measure can be employed within a habitat, in comparison between habitats, or in total.[11] Total biodiversity suffered for tens of millions of years after each of these extinctions. The number of species only returned to pre-extinction levels 25 million years after the Ordovician, 30 million after the Denovian, 100 million after the combination of the Permian and Triassic, and 20 million after the Cretaceous.

We cannot even begin to speculate what life might have been like absent one of the mass extinctions. And it strains our imagination to consider the condition of the Earth if the Permian extinction had been complete. The larger organisms of the planet, what biologist Edward O. Wilson calls "the visible superstructures of energy and biomass pyramids," rely for their existence on biodiversity.[12] Any of the larger species of mammals, including of course *Homo sapiens*, exist only because the mass extinctions were no less and no more complete. Wilson says of insects, were they to completely disappear, "humanity probably could not last more than a few months."[13]

Nobody knows how many species currently inhabit the Earth. One estimate using the tools of systematic biology holds that there is somewhere around 1.4 million species of plants, animals, and microorganisms, over half of which are insects. But even the source of the estimate admits that it "could be off by a hundred thousand, so poorly defined are species in some groups of organisms and so chaotically organized is the literature of diversity in general."[14] The estimate is in any case highly controversial. "[E]volutionary biologists are generally agreed that this estimate is less than a tenth of the number that actually lives on the Earth."[15] By some estimates there are 20–40 million

---

[9] Edward O. Wilson, *The Diversity of Life* (Cambridge, MA: Harvard University Press, 2010), p. 29.
[10] *Ibid.*, p. 30.
[11] Bryan G. Norton, "On the Inherent Danger of Undervaluing Species," in Bryan G. Norton, ed. *The Preservation of Species* (Princeton, NJ: Princeton University Press, 1986), p. 112. See also Wilson, *Diversity*, p. 150.
[12] Wilson, *Diversity*, p. 37.
[13] *Ibid.*, p. 133.
[14] *Ibid.*
[15] *Ibid.*

anthropods, insects, spiders, crustaceans, centipedes, and related species in the tropical rain forests alone.[16] When scientific estimates range from 1.4 million to 100 million and far beyond, it is safe to assume that these are little more than educated guesses.

> To plumb the depth of our ignorance, consider that there are millions of insect species still unstudied, most or all of which harbor specialized bacteria. There are millions of other invertebrate species from corals to crustaceans to starfish, in [a] similar state . . . Consider further that bacteria can evolve rapidly . . . Different strains and even species readily exchange genes, especially during periods of food shortage and other forms of environmental stress.[17]

Our ignorance about the number of species existing produces ignorance about their extinction. The number of unknown species passing into oblivion each year is unknown. But from the species known to scientists, we can be confident that species extinction is proceeding at a rapid pace. One-fifth of the known species of birds around the world have become extinct in the last 2,000 years. About the same percent of freshwater fish have either become extinct or are threatened with extinction.[18] Extending the baseline a bit further back, about 12,000 years ago, after humans arrived in North America, 73 percent of the large mammal genera went extinct.[19] (A genus is a biological classification one step above the species.) Similar rates of extinction occurred with the arrival of humans in Australia, New Zealand, and Madagascar.[20] Historically, extinction has been driven by overkill, habitat destruction, and the introduction of exotic species.[21] Rainforests are bastions of biodiversity on land, but they now cover less than half of their prehistoric area. By some estimates, habitat destruction, through the progressive cutting of the forests, results in loss of about a half a percent of the total species each year.[22]

With the advent of the anthropocene, habitat destruction has been globalized. The consequence of mean temperature increase is that species everywhere are on the move toward the poles and to higher elevations. One recent study found that species are moving to higher altitudes at the rate of 11 meters (m) per decade and toward the poles at 16.9 kilometers (km) per

---

[16] *Ibid.*, p. 140.
[17] *Ibid.*, p. 145.
[18] *Ibid.*, pp. 255–256.
[19] *Ibid.*, p. 247.
[20] *Ibid.*, p. 252.
[21] *Ibid.*, p. 253.
[22] *Ibid.*, p. 276.

decade.[23] Scientific forecasts about species extinction caused by such movements have focused mostly on the danger to species with nowhere to go, either because they are already at high altitudes and latitudes or because there is no corridor for migration. On the basis of factors like these, the Intergovernmental Panel on Climate Change (IPCC) forecasts in its *Fourth Assessment Report* (AR4) that warming of about 1.5°C degrees in the twenty-first century risks extinction of 30 percent of species and the bleaching of most coral reefs. Projection from recent research suggests that focus solely on those factors underestimates the risks of species extinction. New forms of competition between migrating species will develop and these will produce additional extinctions.[24] Whereas warming is causing movement around the globe, acidification is peculiarly affecting the oceans. Oceans absorb CO_2 as part of the natural carbon cycle; and, with an increased concentration of CO_2 in the atmosphere, more of it is cycling down to them. The result is acidification that threatens shellfish and the skeletons of corals. The latter make up the base of coral reefs, which are underwater treasures of biodiversity. This treasure will be depleted when the reefs bleach. The rate of acidification may be cataclysmic to oceanic biodiversity.[25]

ECONOMIC VALUE

The predominant view in early modern philosophy takes value simply as the common measure of economic exchange. Thomas Hobbes writes, "The value of all things contracted for is measured by the appetite of the contractors; and the just value is that which they are contented to give."[26] Value is not an objective property of the things exchanged, but an expression of the appetites of those who would have the things. Hobbes is expansive in his application of this view. His is a version of value known to contemporary economists as "willingness to pay." He takes this account, at least in part, as a corrective to

[23] See I-Ching Chen, Jane K. Hill, Ralf Ohlemüller, David B. Roy, and Chris D. Thomas, "Rapid Range Shifts of Species Associated with High Levels of Climate Warming," *Science* 333 (2011): 1024–1026. http://www.sciencemag.org/content/333/6045/1024 (accessed March 6, 2012).

[24] Mark C. Urban, Josh J. Tewksbury, and Kimberly S. Sheldon, "On a Collision Course: Competition and Dispersal Differences Create No-Analogue Communities and Cause Extinctions during Climate Change," *Proceedings of the Royal Society of Biological Sciences*, January 4, 2012 published online http://rspb.royalsocietypublishing.org/content/early/2012/01/03/rspb .2011.2367.full (accessed Nov. 30, 2013).

[25] See Bärbel Hönisch, et al., "The Geological Record of Oceanic Acidification," *Science* 335 (2012): 1058–1063. DOI: 10.1126/science.1208277.

[26] Thomas Hobbes, *The Leviathan*, (Indianapolis: Hackett Publishing Co., 1994), p. 94 (Pt. I, chp. xv, sec. 14).

the distortions of self-love. A person's value is not accurately measured by an estimation of one's own worth.

> The *value* or Worth of a man is, as of all other things, his price, that is to say, so much as would be given for the use of his power; and therefore not absolute, but a thing dependent on the need and judgment of another. An able conductor of soldiers is of great price in time of war present or immanent; but in peace not so. A learned and uncorrupt judge is of much worth in time of peace; but not so much in war. And as in other things, so in men, not the seller, but the buyer determines the price. For let many (as most men do) rate themselves as the highest value they can; yet their true value is no more than it is esteemed by others.[27]

We are unreliable judges in our own case in part because of the nature of value, which is dependent on the judgment of the valuer.

Hobbes's view is that because a man's value is a measure of his use to others, it is not, or is not simply, a feature of the man himself. One may be strong, agile, handsome, or intelligent, but these are valuable just insofar as someone judges them valuable in light of his or her needs. Values are not real properties of things because the utility of instruments varies with the person in need. So, valuation is an expression of the appetite of the valuer for something deemed useful. In other words, a valuation is the price one is willing to pay in exchange for an item.

Samuel Pufendorf expresses general, even if qualified, agreement with Hobbes regarding the economic foundation for attributions of value:

> Value is divided into common value and eminent value. Common value is found in things and actions, or services, which enter into commerce because they give us pleasure. Eminent value is seen in money, since it is accepted as virtually containing the value of all goods and services and as providing them with a common measure.[28]

Pufendorf avers, however, that the categories of use and value are not identical. There are "some things which are very useful to human life on which no definite value is understood to be set" because either (1) they cannot be owned, or (2) exchanged, or (3) are mere appendages of other items exchanged, or (4) divine law forbids them entering into commerce. As examples of each of these categories, he cites the air and seas, free persons, sunlight and pretty

---

[27] *Ibid.*, p. 51 (Pt. 1, chp. x, sec. 16).
[28] Samuel Pufendorf, *On the Duty of Man and Citizen*, James Tully, ed. (Cambridge: Cambridge University Press, 1991), p. 93 (Bk. I, chp. 14, sec. 2).

countrysides, and sacred acts.[29] In the first and third categories, Pufendorf seems to be offering an early identification of what contemporary economists call "nonexclusive" and "non-rival goods."

The economic value of species and the loss of economic value associated with biodiversity loss are immense. The natural world is a vast and ancient laboratory of chemical innovation, mostly unexploited for human use. The chemical structure of millions of still-existing species has numerous known – and countless unknown – uses that can improve human well-being. Wilson reminds us that,

> Organisms are superb chemists. In a sense they are collectively better than all the world's chemists at synthesizing organic molecules of practical use. Through millions of generations each kind of plant, animal, and microorganism has experimented with chemical substances to meet its special needs. Each species has experienced astronomical numbers of mutation and generic recombinations affecting its biochemical machinery.[30]

We do not have a proper inventory of the chemical compounds that biodiversity constitutes and we have only barely tapped its resources. Even so, 25 percent of all prescriptions issued in the United States are based on compounds extracted from plants, 13 percent from microorganisms, and 3 percent from animals. But currently, only 3 percent of the known flowering plant species have even been scientifically examined for medical purposes.[31]

The untapped nutritional potential of plants is probably enormous. Humans have grown and collected only about 7,000 of the 248,000 known species of plants for food. A mere 20 species provide 90 percent of the world's food; wheat, maize, and rice alone provide more than half.[32] There is similar potential in the animal kingdom. Of the more than 1 million known animal species, we have hunted and ranched just a small fraction. Ranching has been limited primarily to the first domesticated animals.[33] But we are remarkably ignorant about the nutritional value of most plants and animals and the ways in which they might be hybridized for nutrition and gastronomical pleasure.

The economic value of biodiversity is not limited to pharmacological and nutritional goods. Natural ecosystems provide valuable services to human life. Microorganisms create the soil that we need for farming. Insects make plant life possible. Plants absorb CO_2 from the atmosphere during photosynthesis. They

---

[29] *Ibid.*, pp. 94–95 (Bk. I, chp. 14, sec. 3).
[30] Wilson, *Diversity*, p. 285.
[31] *Ibid.*, pp. 285–286.
[32] *Ibid.*, pp. 287–288.
[33] *Ibid.*, p. 294.

also make animal life possible. Natural regions are sources of recreation for us. And within valuable ecosystems, some species play especially important roles. Keystone species are necessary for the flourishing of many other species in the system. Some species exists in a form of mutualism with just one other species. And there are some sets of species that can exist together only according to certain sequences of introduction, which ecologists codify as assembly rules.[34]

There is much uncertainty surrounding any estimation of the economic value of the goods and services associated with the total planetary biodiversity. Indeed there is uncertainty of at least four different kinds. First, we are uncertain of the value to us of the goods and services of all known species in meeting present needs. Second, we are necessarily uncertain of the of the value of the goods and services of known species in meeting needs that will only arise in the future. Third, because we are uncertain about the number and types of species that exist, we are uncertain of the present value to us of unknown species. And fourth, we are necessarily uncertain of the value of the goods and services of unknown species in meeting needs that will only arise in the future. The first and third categories of uncertainty I call "epistemic." In principle, with the extension of possible knowledge about the number and type of species and their use as goods and services in meeting human needs, we could overcome such uncertainty. The second and fourth kinds of uncertainty I call "metaphysical." Because time's arrow points only forward, we cannot know the needs that we will have in the future that we do not have now; we can only conjecture. And we cannot, therefore, know the ways in which existing species might satisfy those needs.

Broadly speaking, there are currently two methodological approaches to estimating the economic value of total diversity. One holds that in weighing the benefits of activities that will destroy natural habitats and render species extinct against the inventory of the known costs, measured in terms of the economic value of the species lost, the benefit of any doubt should go toward preservation. The idea is that uncertainty about how known species might meet present and future needs allows putting a thumb on the preservation side of the scale when the weights are otherwise nearly equal. The other approach holds that for any known species, a presumption in favor of preservation should be maintained until the opportunity costs of preservation become extraordinarily high.[35] The central idea behind this approach is that the costs of extinction,

---

[34] Keystone species, assembly rule, and mutualism are discussed in Wilson, *Diversity*, 163–182.

[35] These two approaches are often distinguished as Resources for the Future (RFF) and Safe Minimum Standard (SMS) respectively. See Alan Randall, "Human Preferences, Economics,

in terms of the direct loss of known and unknown uses of the species and the services that the species provides in maintaining biodiversity, might be very high. This second approach is a kind of precautionary approach to the valuation of known species. Precautionary reasoning is discussed in more detail in Chapter 3.

Although both of these approaches are responses to the uncertainty of the value of species known to us presently, neither seems particularly well suited for the accounting of the unknown future needs that a species would be useful in satisfying. Neither giving the benefit of the doubt to a known species nor establishing a presumption in favor of preservation of a known species can account for the value of unknown species. But the possibility, and sometimes the high probability, that the existence of a known species is necessary for the existence of other unknown ones (because of composition rules, mutualism, or a known species's role as a keystone species) is an additional reason to favor a precautionary approach of assuming the high value of each existing species.[36] The higher valuation of known species in that way indirectly reflects the value to unknown species.

The philosopher Elliot Sober is critical of the precautionary approach in estimating the economic value of species. "If you are completely ignorant of values, then you are incapable of making a rational decision, either for or against preserving some species. The fact that you do not know the value of a species, by itself, cannot count as a reason for wanting one thing rather than another to happen to it."[37] This is, however, an inaccurate representation of the argument for precaution in valuing species. The presumptive high value of any species is not based on sheer ignorance of the value of the species. Rather, it is based on several assertions: (1) Generally, we depend on natural species for nutrition, health, recreation, and other highly important needs; (2) our knowledge of the uses of known species in meeting these needs is incomplete; (3) some species have proved exceedingly useful in this regard; (4) our knowledge of the numbers and types of species is staggeringly incomplete; and (5) our knowledge of the causal relations between species is incomplete. In light of these claims, there is nothing untoward about presuming that the value of any particular species is high in virtue of it serving as a good or service in either meeting a present or future need or in supporting the existence of a known or unknown species that meets (or will meet) such a need.

---

and Preservation" in Bryan G. Norton, ed. *The Preservation of Species* (Princeton, NJ: Princeton University Press, 1986), pp. 79–109.

[36] See also Norton, "On the Inherent Danger of Undervaluing Species."

[37] Elliott Sober, "Philosophical Problems for Environmentalism," in Bryan G. Norton, ed. *The Preservation of Species* (Princeton, NJ: Princeton University Press, 1986), p. 175.

Another criticism of the suitability of attaching economic value to species individually – and biodiversity more generally – contends that the marginal price of any given species is quite low. The philosopher Mark Sagoff contends that for this reason, no conservation argument can be made on grounds of the high-opportunity costs of rendering a species extinct.

> Worldwide the variety of biodiversity is effectively infinite; the myriad species of plants and animals, not to mention microbes that are probably more important, apparently exceed our ability to count or identify them. The "next" or "incremental" thousand species taken at random would not fetch a market price because another thousand are immediately available, and another thousand are immediately available, and another thousand after that.[38]

The argument works only if the market prices for species accurately represent the opportunity costs of rendering them extinct. But cases where there are massive informational problems – cases of epistemic and metaphysical uncertainty regarding the utility of a species in meeting present and future needs – are textbook examples of market failures. We do not have good reasons to trust markets to deliver reliable information regarding opportunity costs if market agents are not sufficiently informed about the goods and services on the market. So, Sagoff's argument simply ignores the problems of extensive uncertainty.

Taking uncertainty into consideration in the course of placing a high presumptive value on any given species is not, however, without its costs to public policy methods. There seems to be no nonarbitrary way of setting the value of any known species according to the precautionary approach. Uncertainty is, after all, uncertainty about the value of the species. If that is the case, although a cost analysis of preserving the species can sometimes be run, there is no clear account of the benefits of preservation to which we can compare the costs. In other words, a well-supported account of how high the opportunity costs of preservation must be before the case for preservation is defeated is unavailable. The argument that species have high-presumptive value on grounds of uncertainty entails that a cost-benefit analysis for purposes of deciding policy regarding species preservation cannot be conducted. In response to this, one cannot help oneself to the utility of cost-benefit analyses as a reason to reject the precautionary approach because in the presence of uncertainty, the usefulness of cost-benefit analyses is called into question.

---

[38] Mark Sagoff, *The Economy of the Earth: Philosophy, Law, and the Environment*, 2nd ed. (Cambridge: Cambridge University Press, 2008), p. 105.

In light only of the economic account of value, what is bad about mass extinction is that it constitutes a presumptively massive loss of goods and services that otherwise would be available for improving human well-being. The importance of this as a general reason sufficient for preserving species requires that the losses to human development caused by a conservation policy be less than those caused by mass extinction. Making the case that mass extinction is wrong therefore requires more than that it is merely bad. It depends on alternative policy not being more costly. This echoes a lesson of Chapter 1: The identification of danger in policy matters depends on there being an alternative policy that is less threatening.

A VERY SHORT EXCURSION THROUGH PUFENDORF AND KANT

A proper appreciation of the use value of species to humanity can go a long way toward explaining why we care about them and the loss of value we incur when we diminish biodiversity. But can it possibly be right to limit the value of species merely to their economic value; to their utility to us? Both the Convention on Biodiversity and the Endangered Species Act express a plurality of values regarding biodiversity and species, with economic value only one among many.

That the economic value is in general too limited to account for all of what we value was already widely appreciated by the seventeenth and eighteenth centuries. By his own lights, Pufendorf's account of value as economic value fails to capture the way that we value ourselves and demand that others value us.

> Man is an animal which is not only intensely interested in its own preservation but also possesses a native and delicate sense of its own value. To detract from that causes no less alarm than harm to body or goods. In the very name of man a certain dignity is felt to lie, so that the ultimate and most effective rebuttal of insolence and insults from others is 'Look, I am not a dog, but a man as well as yourself.'[39]

The need to expand our valuations to include human dignity in addition to price is then already well understood before Kant famously claims that

> In the kingdom of ends everything has either a *price* or a *dignity*. What has a price can be replaced by something else as its *equivalent*; what on the other hand is above all price and therefore admits of no equivalent has a dignity.
>
> What is related to general human inclinations and needs has a *market price*; that which, even without presupposing a need, conforms with a certain taste,

---

[39] Pufendorf, *On the Duty of Man*, p. 61 (Bk. I, chp. 7, sec. 1).

that is, with a delight in the mere purposeless play of our mental powers, has a *fancy price*; but that which constitutes the condition under which alone something can be an end in itself has not merely a relative worth, that is, a price, but an inner worth, that is, *dignity*.[40]

Kant takes the value of skill and diligence to be expressed in a market price, and the value of wit, lively imagination, and humor in fancy price. Price expresses the relative worth of something valued, the worth of it to others. But "fidelity in promises and benevolence from basic principles (not from instinct) have an inner worth,"[41] a dignity.

In the language of this quotation, fidelity in promises and principled benevolence recognize "the condition under which alone something can be an end in itself." Appropriate moral regard of another person is a valuation that is distinct from deeming the individual as useful for our purposes. Dignity is a purpose-independent moral status of persons that requires actions of fidelity and principled benevolence. Moreover, Kant maintains that actions such as these redound to the agent for they "present the will that practices them as the object of an immediate respect."[42] Our capacity to act on maxims based on the dignity of other persons is the source of our self-respect. In general, according to Kant, it is the capacity for such action that makes the attitude of respect appropriate to all – and only – persons.

The conception of dangerous climate change based on the antipoverty principle set out in Chapter 1 rests on a conception of inherent human dignity. The chief interest that the accounts of Pufendorf and Kant have for us in this chapter is not directly in relation to their views on human dignity per se, but in relation to the widening of the kinds of objects of value that they introduce. In this moral inventory, price and dignity are exhaustive of the status of valuable objects. Desire, delight, and their kin are the attitudes appropriate to items valued in terms of price, and respect is the attitude appropriate only to persons valued in terms of dignity. Everything else, including nature, is valued in terms of price.

RESPECT FOR NATURE

There is a broad current of dissatisfaction with valuing everything other than humans merely in terms of price. Karl Marx expresses it: "The view of nature that has obtained under the domination of private property and money is the

---

[40] Immanuel Kant, *Groundwork of the Metaphysics of Morals*, trans. and ed. by Mary Gregor (Cambridge and New York: Cambridge University Press, 1998), pp. 42–43.
[41] *Ibid.*
[42] *Ibid.*

actual despising of and degrading of nature."[43] More recently, certain philosophers have sought to extend the Kantian conception of respect. They agree with Kant that respect is the sole attitude expressive of purpose-independent valuations and apply it promiscuously to valuations of animals and nature generally.

Paul W. Taylor's account of respect for nature is the most thoroughly explicated and trenchantly defended of these revisionist Kantian accounts. Central to Taylor's account is what he calls the biocentric outlook. Biocentrism comprises the following four theses: (1) Humans are members of Earth's community of life, (2) all species are integral elements of a system of natural interdependence, (3) humans are not inherently superior to other species, and, (4) organisms are teleological centers of life.[44] Taylor explains the latter notion as follows:

> To say it is a teleological center of life is to say that its internal functioning as well as its external activities are all goal-oriented, having the constant tendency to maintain the organism's existence through time and to enable it successfully to perform those biological operations whereby it reproduces its kind and continually adapts to changing environmental events and conditions.[45]

Regardless of whether it is a conscious animal or a microorganism, every individual organism's systematic pursuit of its individual good gives it a "unique point of view."[46] This distinguishes organisms from inanimate natural objects, which pursue no goods, and complex machines, which pursue ends derivative of their human makers.

By Taylor's reckoning, biocentrism supports the moral view that organisms have inherent worth, which he takes as meaning that the state of affairs in which its good is realized is better than an otherwise similar one in which it is not.[47] The attitude appropriate to such inherent worth is respect, which involves – among other things – that we have prima facie duties "as stringent as to our fellow humans"[48] of avoiding "doing harm to or interfering with the natural status of wild living things . . . and . . . preserving their natural existence as part of the order of nature."[49] Taylor seeks to absorb the attitudes appropriate to a purpose-independent valuation of organisms into a broadly Kantian account

[43] Karl Marx, *On The Jewish Question* in David McLellan ed., *Karl Marx: Selected Writings* (Oxford: Oxford University Press, 1977), p. 60.

[44] Paul W. Taylor, *Respect for Nature* (Princeton, NJ: Princeton University Press, 2001), p. 99.

[45] *Ibid.*, pp. 121–122.

[46] *Ibid.*, p. 123.

[47] *Ibid.*, p. 75.

[48] *Ibid.*, p. 152.

[49] *Ibid.*, p. 81.

of respect. Expression of respect requires that human interference with nature "be done for the sake of the animals and plants living in a natural state and not for the benefit of humans alone."[50]

There are, however, three serious problems with Taylor's account that render it inadequate for an account of the plurality of ways in which we value species and biodiversity. First, there is a logical gap between the claim that an organism has a good of its own and the claim that we have any sort of duty toward it. That *Mycobacterium tuberculosis* has a good does not entail that any of us should help it pursue its good or even refrain from hindering it. I call this *the normative gap*. Taylor is aware of the normative gap: "One can acknowledge that an animal or plant has a good of its own and yet, consistently with this acknowledgment, deny that moral agents have duty to promote or protect its good or even to refrain from harming it."[51] But he provides no account that would bridge it.

Of the other three theses comprising the biocentric view, two are empirical. One asserts human membership in the Earth's community of life, and the other claims that there is a natural interdependence of all species. Neither of these will bridge the normative gap between recognizing the good of an organism and the claim that we have a duty to promote or protect it.

What about the claim that humans are not inherently superior to other natural organisms? The first thing to say about this is that it is not obvious what the claim is supposed to reject. When we claim that one person or thing is superior to another, we draw, at least implicitly, on certain standards of adequacy or excellence. This includes such matters as the utility of a tool, the beauty of a painting, the clarity of an argument, the accuracy of a quarterback, or the strategic vision of a general. Claims of superiority imply a common coin. We cannot sensibly assert that Barak Obama is superior to Merle Haggard's "Swinging Doors." Claims of superiority also imply that the property compared is scalar. One double India Pale Ale might have a more complex floral flavor than another. In contrast, according to Kant, human dignity is not scalar. One person cannot have more dignity than another.

Taylor's rejection of inherent human superiority seems to commit him to the view that the weight of the claims that organisms make on us in virtue of their purpose-independent value or status is no less than the weight of the claims that humans make on us. For this to be the case, the weightiness of each set of claims would have to be both commensurable and scalar. But if this is the correct way to interpret the third thesis of biocentrism, it does not help

[50] *Ibid.*, p. 94.
[51] *Ibid.*, p. 72.

us to understand how to cross the chasm from the good of the organism to the duties that we have to promote, protect, and refrain from harming it. It simply assumes that the bridge has been built and walked, that the organism's good makes claims on us. The thesis asserts that these claims are no less weighty than the claims humans make on us.

The normative gap arises in any moral theory that builds an account of moral duties out of an account of the impersonal good. There must be an account of why the good is something that we *ought* to promote. John Stuart Mill sees no option but to equivocate: "[T]he sole evidence it is possible to produce that anything is desirable, is that people do actually desire it."[52] According to him, happiness is desirable (it is what we have reason to pursue) because we desire it (we, in fact, do pursue it). Suppose we grant Mill that the fact that we desire something is some kind of reason in support of a duty to promote it. Doubt would still arise when the first-person plural pronoun is not used in characterizing the good valued. Consider the following claim: The organism's good is desirable (it is what *we* have a reason to promote) because the organism desires it (the organism pursues it). If "desirable" is understood as explicated in the parentheses, then the claim is a fairly obvious non sequitur; that others, even great philosophers, have also stumbled where Taylor falls is evidence of the difficulty of the philosophical problem of the normative gap – but not that Taylor's account is satisfactory.

The second problem with Taylor's account is with its insistence that if organisms are the objects of purpose-independent valuation, respect must be the attitude appropriate to that valuation. This invokes a Pufendorfian-Kantian dualism of valuations in which all organisms are assigned a place along with humans as worthy of respect in order to save them from the vagaries of price. As important as it is to appreciate that there is more to valuations than economic value, it is implausible that all purpose-independent valuations are appropriately expressed by the attitude of respect.[53] The respect that we owe other persons is distinct from the concern that we appropriately show our pets. Respect is also distinct from the awe that we naturally experience in the midst of an old-growth redwood forest, the care that is taken in preserving one of Vermeer's paintings, or the wonder that accompanies viewing microbiological

---

[52] John Stuart Mill, *Utilitarianism* in J. M. Robson ed., *The Collected Works of John Stuart Mill, Volume X – Essays on Ethics, Religion, and Society* (Toronto: The University of Toronto Press, 2006), p. 282. http://files.libertyfund.org/files/241/Mill_0223--10_EBk_v6.0.pdf (accessed March 30, 2012).

[53] The first chapter of Elizabeth Anderson's, *Value in Ethics and Economics* (Cambridge, MA: Harvard University Press, 1993) is useful in laying out the limitations of value monism and dualism. The pluralism that I defend in this chapter owes much to Anderson's account.

activity. Concern, awe, care, and wonder – although distinct from the attitude of respect – all suggest that their objects possess a status not captured by mere price. The universe of value contains many treasures, not merely dignity and price.

Finally, Taylor's account is of little use in helping us understand our valuation of species and biodiversity because it is, at base, individualistic. "[U]nless individuals had a good of their own that deserves the moral consideration of agents, no account of the organic system of nature-as-a-whole can explain why moral agents have a duty to preserve *its* good."[54] There is no clear path that leads to a derivative moral concern for organic systems, including species and their diversity, from an individualist starting point. For Taylor, the polar bear sitting on the ice is the object of respect; *Ursus maritimus* is not, nor is the diversity of species in the Arctic. But insofar as we value biodiversity, we are not valuing that bear. His death is not what we seek to avert. Preserving biodiversity is directed toward his species, not toward helping him.

THE LAND ETHIC

Aldo Leopold's land ethic includes an attempt at an alternative to economic valuations of species. After raising doubts about the adequacy of considering songbirds as merely economically valuable, he extends the account to predatory species.

> A parallel situation exists in respect of predatory mammals, raptorial birds, and fish-eating birds. Time was when biologists somewhat overworked the evidence that these creatures preserve the health of game by killing weaklings or that they control rodents for the farmer, or that they prey only on 'worthless' species. Here again, the evidence had to be economic in order to be valid. It is only in recent years that we hear the more honest argument that predators are members of the community, and that no special interest has the right to exterminate them for the sake of a benefit, real or fancied, to itself.[55]

Here the value of a predatory species derives from the role that it plays in a larger system of interdependence. This sort of value might be called "ecological value."

Leopold refers to the larger system of interdependence as "the biotic community," which is perhaps a bit narrower than the term "ecosystem" since the latter possesses abiotic elements as well as organisms. But nothing seems

---

54 Taylor, *Respect for Nature*, pp. 118–119. Emphasis in original.
55 Aldo Leopold, "The Land Ethic" in *The Sand County Almanac*. http://home.btconnect.com/tipiglen/landethic.html (accessed March 30, 2012).

to be lost by bringing his view up to date as enjoining the preservation of the integrity, stability, and beauty of ecosystems. According to this view, the ecological value of a species seems to depend ultimately on the value of the ecosystem of which the species is a part. Leopold advocates a value pluralism that requires us to "quit thinking about decent land-use as solely an economic problem. Examine each question in terms of what is ethically and esthetically right, as well as what is economically expedient." He claims that, "[a] thing is right when it tends to preserve the integrity, stability, and beauty of the biotic community. It is wrong when it tends otherwise."[56] Moreover, Leopold identifies the attitudes of love, respect, and admiration as appropriate to the value of land understood as a system of biotic communities.

There are two attractive features of Leopold's view. He offers an account of the object of valuation that is nonindividualistic; it is not reducible to the value of individual organisms. Instead, it is biotic communities or ecosystems that are valuable. Species are valuable for their ecological roles within the ecosystem. And he recognizes several kinds of value: economic, ecological, aesthetic, and moral. The view rests ultimately on the idea that ecosystems possess value, in virtue of which species are ecologically valuable. But what sort of value could that be?

Holmes Rolston III attempts an account of the value of ecosystems. He claims that the ability to value – or what he calls "value-ability" – is the source of value.[57] Value is projected by valuers. Moreover, valuers take some things as instrumentally valuable for the sake of other things. Some of the other things are valued non-instrumentally. Chief among these are the valuers themselves. Insofar as that is the case, valuers are non-instrumentally valuable. For example, Rolston contends that species and ecosystems are valuers; they value themselves for their own sake. Hence, they are non-instrumentally valuable.

Let's consider this more carefully. Rolston asserts that "[a] valuer is an entity able to defend value."[58] Such defense does not require consciousness or sentience. The defense systems that organisms have against predation count as defending their own value. Building on this, he claims the following: "The single, organismic-directed course is part of a bigger picture in which a species too runs a telic course through the environment, using individuals resourcefully to maintain its course over much longer periods of time . . . The species defends a particular form of life, pursuing a pathway through the world, resisting death

[56] *Ibid.*
[57] Holmes Rolston III, "Value in Nature and the Nature of Value," in Robin Attfield and Andrew Belsey, eds., *Philosophy and the Natural Environment* (Cambridge: Cambridge University Press, 1994), pp. 13–30.
[58] *Ibid.*, p. 18.

(extinction), by regeneration maintaining a normative identity over time."[59] Appreciating valuing as an activity of attributing value to something requires using verbs in the active voice. Rolston uses similar language for ecosystems. "Ecosystems are selective systems, as surely as organisms are selective systems. The system selects over the long ranges for individuality, for diversity, for adapted fitness, for quantity and quality of life."[60] Now, in evolutionary biology terms it is highly controversial, to say the least, to attribute selection activity to the ecosystem. But the point is not whether Rolston is correct in making that particular claim, but to see how he arrives at the view that ecosystems are non-instrumentally valuable.

Even if we grant Rolston his descriptions full of the active voice valuing activity of species and ecosystems, we are nonetheless confronted with a familiar problem. From the claim that species and ecosystems value themselves, there is no reason for concluding that we have reason to value them. The normative gap opens up again. That X has a good of its own or values itself does not establish that we have reason to value X. It does not take a lot of imagination to substitute particulars for X that make the point plain.

As we have seen, the individualism of duties to organisms will not provide the right kind of reasons for valuing species. And it is very hard to make out how species or biodiversity could be the objects of direct moral duties. This suggests that there may be more promise in Leopold's invocation of aesthetic value.

THE AESTHETIC VALUE OF ORGANISMS

We might expect those who have devoted their lives to the study of art to have a more subtle appreciation of aesthetic qualities than those of us less tutored. It is remarkable good fortune, then, for those of us seeking to better understand the aesthetic qualities of nature that John C. Van Dyke, an accomplished art critic and professor of Art History at Rutgers University, lived to publish the journal of his three-year-long desert wanderings. In 1898 he packed a few provisions onto a horse and, with his fox terrier, wandered into the Coachella Valley through the San Gorgonio Pass, cut by the great San Andreas Fault between the San Bernardino Mountains on the north and the San Jacinto Mountains on the south. Palm Springs was not the resort mecca that it now is and agriculture was only beginning to arrive in the Imperial Valley. The vast arid area between the Colorado River, the Laguna Mountains, the Mexican

[59] *Ibid.*, p. 21.
[60] *Ibid.*, p. 24.

border, and the Mojave Desert was then commonly known as "the Colorado Desert."

In the Colorado Desert, Van Dyke records his impressions of the landscape, the air, the sunlight, and desert flora and fauna. With an eye well-trained from years of intense study of art, he is witness to a world rich in aesthetic value. He is keenly aware of the perceptual qualities of sunlight at various times of the day and, remarkably, even of the color of the air at different elevations. And when he discusses the flora and fauna, it is the properties of species that capture his attention, not the well-being of the individuals.

> [N]ature shows her absolute indifference to the life or death of the individual. She allows the bugs and beetles to be slaughtered like the mackerel in the sea. But she is a little more careful about preserving the species . . . Thousands are annually slaughtered; yes, but thousands are annually bred. What matter about their lives or deaths provided they do not increase or decrease as a species![61]

By Van Dykes's estimation, desert plant life possesses expressive properties that are the object of appreciation. The tenacity of desert plant life is the material of high drama complete with struggle, resolution, and hope. "Never by day or by night do they lose the armor or drop the spear point. And yet with all the struggles they serenely blossom in season, perpetuate their kinds, and hand down the struggle to the newer generation with not jot of vigor abated, not title of hope dissipated."[62]

There is also a delightful harmony between the features of a desert plant and its environment:

> Are they beautiful these plants and shrubs of the desert? Now just what do you mean by that word "beautiful"? Do you mean something regular of form, something smooth and pretty? Are you dragging into nature some remembrances of classic art; and are you looking for the Dionysius face, the Doryphorus form, among these trees and bushes? If so the desert will furnish you not too much beauty. But if you mean something that has a distinct character, something appropriate to its setting, something admirably fitted to a design (as in art the peasants of Millet or the burghers of Rembrandt and Rodin), then the desert will show forth much that people nowadays are beginning to think beautiful. Mind you, perfect form and perfect color are not to be despised; neither shall you despise perfect fitness and perfect character.[63]

---

[61] John C. Van Dyke, *The Desert* (New York: Peregrine Smith, Inc., 1980), p. 190.
[62] *Ibid.*, p. 149.
[63] *Ibid.*, p. 144.

Van Dyke also emphasizes the perceptual properties of desert plants. The adaptation of a plant species to the desert environment produces instances with subtle color harmonies that are pleasing to behold. "The desert plants, every one of them have very positive characters; and I am not certain but that many of them are interesting and beautiful in form and color."[64]

The adaptations that produce these positive aesthetic responses are not, of course, peculiar to plants. Van Dyke is also much impressed with the animal species of the desert. Here as well are expressive qualities reminiscent of drama. The theme is mortality, and the genre is the chase:

> Everything pursues or is pursued. Every muscle is strung to the highest tension. The bounding deer must get away; the swift-following wolf must not let him. The gray lizard dashes for a ledge of rock like a flash of light; but the bayonet bill of the road runner must catch him before he gets there. Neither can afford to miss his mark. And that is perhaps the reason why there is so much development in the special directions, so much fitness for a particular purpose, so much equipment for the doing or the avoiding of death.[65]

Van Dyke also notices the perceptual qualities of desert animal life that are the bases of positive aesthetic judgment.

> Even the classic idea of beauty, which regards only the graceful in form or movement or the sensuous in color, finds types among these desert inhabitants. The dullest person in the arts could not but see fine form and proportion in the panther, graceful movement in the antelope, and charm of color in all the pretty rock squirrels.[66]

Natural adaptations produce form and color that are pleasing to behold.

Van Dyke claims that natural beauty is a factor in the beauty of artistic representations of nature.

> These animals have made the best of the worst, and their struggle has given them a physical character which is, shall we not say, beautiful? Perhaps you shudder at the thought of a panther dragging down a deer – one enormous paw over the deer's muzzle, one on his neck, and the strains of all the back muscles coming into play . . . Look at the same subject done in bronze by Barye and you will see what a revelation of character the great statuary thought it. Look, too, at Barye's wolf and fox, look at the lions of Géricault,

---

[64] *Ibid.*
[65] *Ibid.*, pp. 155–156.
[66] *Ibid.*, pp. 171–172.

and the tigers and serpents of Delacroix; and with all the jaw and poison of them how beautiful they are.[67]

The artist sees something in the animal that makes it worthy of representation. That property or those properties are as fit for aesthetic evaluation in the subject as they are in the work of art. There are obvious cases – the panther's form, the antelope's grace, and the rock squirrel's charm of color – of appropriate aesthetic evaluation in representatives of animal species. As a general rule, however, it would be fallacious to infer the beauty of the represented from the beauty of an artistic representation. But when we have experiences of organisms as beautiful, that they can also be represented beautifully in forms of art provides a reason to believe that natural beauty is one of the factors in the beauty of the artistic representation of nature.[68] Van Dyke, however, avoids the implausible claim that has tempted others – namely, that artistic representations of nature have no aesthetic factors other than the beauty of the nature they represent.[69]

Aesthetic evaluation applies to the individual organism with which we are acquainted, but the natural properties that make it appropriate for evaluation are those that it has as a member of its kind or species as the result of natural selection.[70] The gracefulness of the individual antelope is owing to the properties that it possesses as a member of the species. It is in virtue of an organism being a good instance of its kind that we find it aesthetically pleasing. This allows us to understand why we value members of rare species more highly than members of non-rare species. The former are more highly valued because the opportunities for appreciating them in their natural setting are fewer and the likelihood that we will lose all opportunities for viewing them is higher. For the loss of a species is analogous to the loss of a genre of art.[71]

Moreover, given Van Dyke's account of the interplay of species with their living and nonliving environments, members of species possess the natural features that are the objects of aesthetic evaluation, in part, because of the

---

[67] *Ibid.*

[68] I am grateful to Diarmuid Costello for a discussion of these issues.

[69] For apparent instances of the implausible claim, see the following: "Art cannot rival this pomp of purple and gold. Indeed the river is a perpetual gala, and boasts each month a new ornament" (Emerson, *Nature*, sec. 3). And, "the hardly esoteric judgment that paintings of the Matterhorn and purple heather are kitsch has a scope reaching far beyond the displayed subject matter. What is innervated in the response is, unequivocally, that natural beauty cannot be copied" (Theodor Adorno, *Aesthetic Theory*, Robert Hullot–Kentor, trans. and ed. [London and New York: Continuum, 1997], p. 67).

[70] See also Lilly Marlene-Russow, "Why Do Species Matter?" *Environmental Ethics* 3 (1981): 101–112.

[71] This is convincingly argued in Alan Carter, "Biodiversity and All the Jazz," *Philosophy and Phenomenological Research* LXXX (2010): 58–75.

interaction of a species with other species. The interaction that Van Dyke emphasizes is predation, almost to the total exclusion of mutualism. This is either exaggeration or oversight, but it suffices for his point that species do not – indeed, cannot – exist without interaction with other species. This is another instance of the ecological value of a species. As with the economic value of species, the aesthetic value of the members of any particular species is strongly dependent on the existence of other species. This interdependence of species increases the disvalue in the extinction of a particular one. Unlike artistic genres, the loss of a species often puts extinction pressure on other species.

Van Dyke seems to affirm the doctrine of the positive aesthetics of nature; that unlike art, nature provides no bases for negative aesthetic judgments.

> [T]here is nothing ugly under the sun, save that which comes from human distortion. Nature's work is all of it good, all of it purposeful, all of it wonderful, all of it beautiful . . . Each in its way is suited to its place, and each in its way has its unique beauty of character. And so, more truly perhaps than Shakespeare himself knew, the toad called ugly and venomous still holds a precious jewel in its head.[72]

The scope of the claim that all of nature is beautiful is vague. Does it apply to every living thing? To every natural thing? To the forest or to the trees? Is it an all-things-considered judgment that allows that the parts of natural things may be not pleasing? Or must each part be beautiful as well? Is the claim to be understood as necessarily true or only contingently so? These questions make assessing the doctrine of positive aesthetics a complicated matter, and the doctrine itself rather controversial.[73] Van Dyke does not clarify his claim. His exuberance about the matter, however, does not affect the central point that I have been discussing – namely, that the beauty of individual organisms, judged at least in part by their being instances of species, provides us with a reason to protect their species and to treasure biodiversity.

VALUING BIODIVERSITY

We are now in a position to see how the four desiderata for an account of the value of biodiversity that I discussed near the beginning of this chapter can be met.

---

[72] Van Dyke, *The Desert*, pp. 192–193.

[73] For an influential philosophical defense of the doctrine, see Alan Carlson, "Nature and Positive Aesthetics," *Environmental Ethics* 6 (1984): 5–34. The various questions I mention about the doctrine are discussed with clarity in Malcolm Budd, *The Aesthetic Appreciation of Nature* (Oxford: Oxford University Press, 2002), pp. 97–109 and 125–148.

First, we can appreciate several ways in which biodiversity is valuable. It is economically valuable because of the many valuable goods and services that organisms from a variety of species provide, or could provide, for us. Moreover, a diversity of species provides us with opportunities for the aesthetic appreciation of individual organisms in their natural setting. The immediate object of perception is an individual plant or animal, but the physical properties of the individual that are the bases of the positive aesthetic evaluation are ones that it has as a member of a species. So, for example, majesty is seen in the particular bighorn sheep seen in Rattlesnake Canyon, but it possesses this because it is an outstanding instance of its kind. Bighorn sheep possess the physical properties that are the object of the positive aesthetic evaluation because of untold generations of adaptation to the environment around it. This includes not only nonliving features of terrain and climate, but prey and predation and various forms of competition and mutualism with other species. Species diversity is also, then, ecologically valuable. Any given species is adapted to an environment that includes other species. Certain aspects of the aesthetic evaluation of the individual member of a species will necessarily make reference, even if only implicitly, to the broader natural context of biodiversity. The flame-like crimson blooms of the ocotillo are more fully appreciated as an element in the larger context, in which they form a signal to pollinators such as hummingbirds and carpenter bees. The aesthetic evaluation of an individual organism involves recognizing it as an exemplary instance of the species, which has acquired its specific characteristics as the result of the interaction of individuals in the broader context of biodiversity.

Second, this account provides us with an understanding of concern, delight, wonder, and awe at nature; an account that is independent of whether or not the attitudes derive ultimately from a religious worldview, which makes it suitable for political purposes. The economic value of species merits our concerned efforts at conservation as a kind of savings plan. And several of the aesthetic properties of desert life that Van Dyke observes, including gracefulness, color, complexity, and systemic integration, are appropriate objects of attitudes of delight, awe, and wonder. Moreover, humility and gratitude are supported by a scientific understanding of the timescale of millions of years, over which current species evolved to populate our planet. When we come to appreciate the evolution of biodiversity over this long period, our natural place in the scheme of things seems small and we may form an attitude of impersonal gratitude; not for a favor intentionally done to us by anyone, but for the existence of these impersonal forces. As the biologist Wilson emphasizes, we literally owe our existence to the diversity of life. "This is the assembly of life that took a billion years to evolve. It has eaten the storms – folded them into it genes – and created the world that created us. It holds us steady."[74] But, of course, concern, delight, awe, wonder, humility, and gratitude are also no strangers in religious understandings of the natural world.

Third, the account makes sense of the value pluralism stated in several legal documents directed toward species conservation. A plurality of value-expressing attitudes and actions are appropriate to objects of both economic value and aesthetic appreciation. We show concern for the conservation of that which we value economically. We delight in the appreciation of artistic beauty. We take care to preserve and restore beautiful things. We devote considerable resources to storing art in secure climate-controlled buildings and to maintaining buildings and monuments that are exposed to the elements. We take these as part of our common heritage and we assume a responsibility to preserve them for future generations to appreciate. We admire the interplay of physical properties that are the bases of our aesthetic appreciation of paintings, sculptures, and music. We make efforts to understand these in order to gain a greater appreciation of the art and, in some cases, to make art ourselves. Sometimes we are in awe of the power of an artwork: its force, its scale, or its grandeur. We are often humbled by the skill or creative processes that produce works of beauty. We admire the talents and skill of the artist. We are grateful for the many ways in which aesthetically pleasing things enrich our lives and we seek such enrichment. We seek to understand how a work fits into a genre and how it is similar to and different from other works in the genre. These familiar responses of delight, care, preservation, admiration, awe, and efforts to understand objects of art constitute a plurality of expressions of our positive valuations of them.

Analogous responses are appropriate to the aesthetic appreciation of nature. This is not to suggest that the aesthetic appreciation of nature is identical to the appreciation of art. There are many important differences.[75] One is that with art, both our aesthetic appreciation and our preservationist concern are with individual works of art, whereas although we appreciate the individual organisms, we seek to preserve the species. Unlike works of art, the individual instances that we experience come to be and die in a natural cycle of life and death. We delight in features that the individual has as a member of a species that extends before and after the life of the individual. We seek to preserve the species and to understand its role in its environment in order to continue and to deepen the positive responses we have when perceiving the individuals. We

---

[74] Wilson, *Diversity*, p. 15.

[75] I am grateful to Martin Seel for a discussion of the many ways in which the aesthetic appreciation of nature differs from the aesthetic appreciation of art.

may be awestruck by understanding that a particular organism is the result of a very lengthy career of adaptation of the species. This plurality of attitudes and actions expresses our evaluation of biodiversity. The case of art shows that we take such evaluations as strongly normative. Appreciation of positive aesthetic value gives us reason to hold a variety of beliefs and to do various things, including devoting significant resources to preservation and restoration.

Fourth, the account explains the value of biodiversity. The economic value of species is explained by the goods and services individual instances of the species render us. Various features of natural organisms are the basis for judgments of their aesthetic value. Species engaged in a pitched struggle to survive in a hostile environment display various expressive values. The physical objects of our positive aesthetic evaluations are strongly context dependent. What we appreciate makes sense within the context of survival, although our appreciation of it is not limited to our understanding it in that way. We may both marvel at the bloom of the ocotillo and become fascinated with its role within the lives of a community of pollinators. And our understanding of the dynamics of that community may result in awe for the complex processes that brought it about and the relative equilibrium between the members that maintain it for significant periods of time.

The aesthetic value of species-members provides us with a source of reasons to motivate the project to preserve biodiversity. This understanding of the aesthetic value of organisms, for example, would support judgments that anthropogenic species extinction is either gross negligence or vandalism. It is akin to someone either failing to take proper care of entire genres of great works of art or wantonly destroying them.[76]

AN OBJECTION AND A RESPONSE

Not everyone, however, agrees that aesthetic value is the right kind of value to underwrite the preservationist project. The counter argument often takes a surprising turn. Rather than discussing very real threats to biodiversity that human development and climate change pose, several philosophers have thought it useful to entertain fanciful examples in which the last person alive on Earth has the power to plan and set into motion the destruction of both all the works of art and all of nature after he dies.[77] We are to imagine

---

[76] See also Robert Elliot, *Faking Nature: The Ethics of Environmental Restoration* (London and New York: Routledge, 1997), pp. 57–58.

[77] Early discussion of this so-called last man argument appear in Stanley I. Benn, "Personal Freedom and Environmental Ethics: The Moral Inequality of Species," in Gray L. Dorsey, ed.,

some sort of delayed process; a long fuse, if you will, lit before he dies that will, after he is dead, destroy the art that still exists or all life that still exists. To make the two possibilities comparable, we have to imagine further that the loss of life would be painless and unannounced, an instant vaporization unaccompanied by any sort of physical or psychological pain to sentient animals. That additional requirement fixes our attention on the destruction of species rather than the painful death of individuals. Several philosophers, after considering examples like this, report that they have a different reaction to the destruction of art and life and that the difference is indicative of a difference in the kind of value of each. Robert Elliot, for example, claims, "The horror that I feel at the destruction of natural value seems to me to be genuinely moral horror."[78] Elliott feels sufficiently confident in his own intuitions to conclude that our valuation of nature is not only aesthetic, but moral.

The thought experiment seems to me to be of little value. Given how different the imaginary situation is from our own and the degree to which such examples leave our responses unconstrained, I have no confidence that the reports of anyone's reactions tells us anything reliable or significant. I see four problems with this approach. The first is that, given how unusual the imaginary case is, it is difficult to know whether our reactions to it are the same as our would-be reactions to a real case like it. Because we are not asked to deduce a conclusion from a premise, but to gauge total attitudinal responses to a very stylized situation, it is surprising that anyone could confidently proclaim that their response to the imaginary case is what their response in the analogous real case would be. Elliott speaks of "genuinely moral horror," but, given the context of the response, it's hard to see how any response, in fact, could be genuine. Consider whether individuals who have never been in combat could confidently proclaim how they would respond in the situation. If we want to know something about the nature of our values, this seems an unreliable test. A second problem arises from generalizing the intuition. Even if, when pressed, Elliot is confident about his intuitions in the imaginary stylized case, surely people's intuitions are bound to differ about these matters, precisely because there is no common experience nearer enough to this example that might help to fix our intuitions.

Equality and Freedom: International and Comparative Jurisprudence, vol. 2 (Dobbs Ferry: Oceana Publications, 1977), pp. 401–424 and in Richard Routley and Val Routley (1979), "Against the Inevitability of Human Chauvinism" in K. E. Goodpaster and K. M. Sayre, eds., *Ethics and the Problems of the 21st Century* (South Bend, IN: Notre Dame University Press, 1979), pp. 36–59.
[78] Elliott, *Faking Nature*, 70.

Third, even if we – or most of us – could be confident that our common reaction to the imaginary case is similar to what our reaction to a real one would be, it is unclear what that would tell us about our valuations in the real world. We are not faced with a one-off vaporization of all species taken in isolation from everything else that we value, but a piece-by-piece degradation of biodiversity in the context of competing values. It is not obvious that isolating our response in the manner of the example is at all helpful for understanding our values in the real world. Finally, even if we could be confident that the thought experiment points to differences in our valuation of art and nature in the real world, it is unclear why we should interpret those as indicative of different kinds of values rather than, say, differences in intensity of the same kind of value. If one finds oneself more horrified at the prospect of the final destruction of all nature than of all art, it would seem reasonable to conclude that one is simply more horrified, not differently horrified. If, following Kant, valuing involves taking "pleasure in loving something without any intention of using it,"[79] presumably anyone who cherishes something as aesthetically valuable would be pained by its destruction regardless of whether destruction were to occur after one could appreciate it.[80]

There are differences in the loss between, on the one hand, negligently letting several great paintings decay and, on the other hand, destroying an ecosystem and driving countless species into extinction. In the latter case, a dynamic and self-reproducing system is destroyed and along with it, natural kinds that have been evolving over millions of years. There will be no more individual instances of the species in the future. The integrity of the ecosystem and the ancient processes that produced it are not artifacts, and the wonder and humility that we experience in appreciating the ecosystem and its constituent species is in part a response to their being natural.[81] In the absence of our influence, the ecosystem and its constituent species would have continued to exist and to evolve naturally, whereas in the absence of our care, individual paintings – but not entire genres – would have been slowly destroyed by the forces of nature. The maintenance of a painting requires insulating it from

---

[79] Immanuel Kant, *The Metaphysics of Morals*, p. 443 and Immanuel Kant, *Ethical Philosophy*, James W. Ellington, trans. (Indianapolis and Cambridge: Hackett Publishing Co., 1983), p. 106.

[80] See also Thomas E. Hill, Jr., "Ideals of Human Excellence and Preserving Natural Environments," *Environmental Ethics* 5 (1983): 211–224.

[81] Robert E. Goodin claims that the importance of naturalness is that it provides a context for us to make sense out of our lives. See Robert E. Goodin, *Green Political Theory* (London: Polity Press, 1992), pp. 30–41. Dale Jamieson claims that nature is respect worthy because it provides the background conditions for our lives to have meaning. See Dale Jamieson, "Climate Change, Responsibility, and Justice," *Science and Engineering Ethics* 16 (2010): 442.

natural processes whereas the maintenance of species can often be best achieved by preserving it in its natural surroundings. But these differences do not warrant the conclusion that the appreciation of art involves aesthetic valuations whereas the appreciation of nature involves moral ones. Moreover, because the appreciation of the latter appeals to features similar to the appreciation of the other, as Van Dyke notices, there is good reason to believe that the kind of valuation is the same.

REGRET AND RESPONSIBILITY

"If the case of natural beauty were pending, dignity would be found culpable for having raised the human animal above the animal"; with that claim, Theodor Adorno suggests that the recognition of human dignity poses a deep threat to natural beauty.[82] There is, to be sure, the possibility of tension. But the pluralistic account of value defended and applied in these first two chapters can accommodate both the value of human dignity and of natural beauty. This pluralism ensures that the latter is not lightly – and not without regret – sacrificed on behalf of the former.

Appreciation for the value of biodiversity includes a plurality of valuations and attitudes. As I discussed earlier in this chapter, biodiversity is appropriately valued not only aesthetically, but also economically. And if ecosystems are aesthetically valuable, species are also presumably ecologically valuable. These considerations augment the pluralism of values at play in appreciating biodiversity and provide us with a manifold of reasons to be concerned about climate change in addition to those based on the effects of climate change on human beings. There is pluralism not only in the valuations of biodiversity, but also between the valuations of biodiversity and human values. It is noteworthy that pluralistic accounts of value offer validation to our rational regrets. Regret is an attitude appropriate when in order to preserve something of value we must lose something else of value.

The aesthetic value of an organism is a kind of impersonal value. It tells us nothing about what we owe to each other.[83] Kant, however, maintains that the disposition to value beautiful things is a character trait "favorable to morals" because valuing beautiful things involves seeing them as something other than mere means to our ends.[84] Aesthetically valuable things have a

---

[82] Adorno, *Aesthetic Theory*, p. 62.

[83] This distinction between impersonal and personal values draws on T. M. Scanlon, *What We Owe to Each Other* (Cambridge, MA: Harvard University Press, 1998), pp. 218–223.

[84] Kant, *Metaphysics of Morals*, 106.

kind of purpose-independent value. Still, there is plenty of elasticity between being disposed to love beauty and acting to preserve it. One who takes aesthetic value seriously can nonetheless permit the destruction of aesthetically valuable objects, if morality requires it.

Whether the various valuations of species serve as good reasons to preserve them depends on the moral-opportunity costs of doing so. For example, the economic benefits of species preservation can, in principle, be outweighed by the costs. The strongest kinds of moral reasons that would make the case against preserving species invoke the antipoverty principle defended in Chapter 1. If there are cases in which species' extinction is necessary in order not to prolong or deepen abject poverty, then the loss of economic, ecological, and aesthetic values are justified on grounds of human dignity.[85]

The moral value of human dignity is exceedingly important to us – more so than aesthetic value. As Pufendorf notes, there is something especially troubling about a person who disregards human dignity. "Human nature . . . belongs equally to all and one would not gladly associate with anyone who does not value him as a man as well as himself and a partner in the same nature."[86] Such a person is a danger to us in a way that one who fails to have appropriate aesthetic valuations is not.[87] Still, the loss of value that occurs when species become extinct can never be erased from the cost column in our comprehensive evaluative ledger. There may be reasons to regret some of the consequences of actions, even if there are important gains that result from doing them. And a standing suspicion about the claim that the losses could not have been avoided seems appropriate. Allowing the excuse of necessity in general is consistent with scrutinizing its every invocation in the court of reasons. Where there is necessity, regret is appropriate; but where necessity is lacking, excuses will be found wanting.

When considering whether some species extinction is unavoidable given the pursuit of human development, it is appropriate to look not only at local actors, but to the international context. In Chapters 5 and 6, I argue that

---

[85] Adorno doubts that we are capable of aesthetic evaluation when so confronted by the demands of natural survival: "Times in which nature confronts man overpoweringly allow no room for natural beauty; as is well known, agricultural occupations, in which nature as it appears is an immediate object of action, allow little appreciation for landscape," *Aesthetic Theory*, p. 65. According to this view, there are no conflicts in values of the kind that I am discussing in this paragraph. Nicole Hassoun also defends an anthropocentric approach that prioritizes human values in "The Anthropocentric Advantage? Environmental Ethics and Climate Change Policy," *Critical Review of International Social and Political Philosophy* 14 (2011): 235–257.

[86] Pufendorf, *On the Duty of Man*, p. 61 (Bk. I, chp. 7, sec. 1).

[87] For a recent defense of Pufendorf's claim, see Scanlon, *What We Owe to Each Other*, pp. 160–168.

meeting the requirements of sustainability is primarily the responsibility of the highly developed countries. When a poor country pursues human development projects that result in biodiversity loss, there might have been alternatives that would have been economically more attractive on the condition that there were appropriate investments from industrialized countries. The ability-to-pay conception of responsibility discussed in Chapter 6 is relevant here. When urgent human development needs are on the line, the preservation of biodiversity is not merely a matter of local responsibility; when those who are better-off act responsibly, regrets can be minimized.

3

Risks, Uncertainties, and Precaution

"It is a world of change in which we live, and a world of uncertainty. We live only by knowing something about the future; while the problems of life, or of conduct at least, arise from the fact that we know so little."

– Frank H. Knight

Climate change poses risks of catastrophic changes to human communities and ecosystems. Left unmitigated, the risks of climate change are dangerous, in particular to persons who will be made especially vulnerable by the conspiracy of poverty and geography: those residing in huge mega deltas in North Africa and East Asia, or living along major glacier-fed rivers in Asia and South America, or engaged in subsistence agriculture in the arid regions of central and southern Africa. Climate change also threatens mass extinctions of species. But one thing undeniably true about the future climate is that we do not have detailed knowledge about how different it will be from our present climate. That truth gives us pause. If we don't know, how can we plan for adaptation? Why should we take action to mitigate climate change? If this question is asked rhetorically, it sounds like a form of climate change skepticism. The claim that we do not know *exactly* what is in store for us if climate change is left unmitigated is not false. But the policy conclusion drawn by climate change skeptics most certainly is.

Science can reliably tell us a great deal about climate change, even if it cannot reliably forecast what the annual rainfall in the American Midwest will be in 100 years. The United Nations Framework Convention on Climate Change echoes the language of the 1992 Rio Declaration in declaring that lack of scientific certainty is no reason to postpone action: "The Parties should take precautionary measures to anticipate, prevent, or minimize climate change and mitigate its adverse effects. Where there are threats of serious or irreversible damage, lack of full scientific certainty should not be used as a reason for postponing such measures."[1] This is a relatively weak claim; uncertainty does not justify regulatory inaction. In this chapter I shall argue for a stronger claim; namely, that various uncertainties regarding climate change support taking precautionary regulatory action. In other words, uncertainty adds to the reasons for mitigation discussed in Chapter 1. The view that uncertainty can warrant precautionary measures has come under attack by some policy analysts. But once the nature of the uncertainty is better understood, such attacks will seem misguided. I begin by clarifying risk and uncertainty generally and in the climate system in particular. I follow that with a discussion of when it is reasonable to take action under conditions of uncertainty and I argue that the possible costs associated with getting it wrong – and the relatively low costs of mitigation policies – support taking action to mitigate climate change even though there is uncertainty about the effects of climate change.

RISK AND UNCERTAINTIES

Most outcomes in life are uncertain in a colloquial sense. When we throw the die, we do not know whether we will get the desired number. When we fly in a plane, we do not know whether we will arrive safely. But in the former case, owing to the nature of the die (assuming it to be a fair one) we know a priori what the likelihood of getting the number will be. In the second case, we can rely on a massive amount of evidence about the incidences of air accidents to know how likely it is that the plane will crash, if one is sufficiently worried. This latter probability can be further refined by looking at the evidence for the type of plane and the carrier in question. These two cases differ insofar as the latter relies on studying our experience and the former doesn't, but in both we speak of the risks of the action – either betting on the roll of the die or flying in a plane.

The study of probability is divided into two camps. There are those who hold that inferences based on probabilities establish grounds for believing in objective relations between independent objects and those who believe all probabilistic inferences are based on applying "habits of the human mind," as Frank Ramsey puts it, to factual beliefs to yield new beliefs, and these habits are to be judged by how useful they are in producing new beliefs.[2] This is not the place to begin to judge this dispute that is fundamental to our understanding

[1] United Nations Framework Convention on Climate Change, Article 3, Paragraph 3. http://unfccc.int/resource/docs/convkp/conveng.pdf (accessed July 4, 2013).

[2] Frank Plumpton Ramsey, "Truth and Probability," in *The Foundations of Mathematics and other Logical Essays* (London: Routledge and Kegan Paul Ltd., 1931), pp. 196–198.

of what probabilistic inference is. It is enough to note that judgments of risk are reliable to the extent that the probability judgments that they are based on are reliable, for we take a risk to be the product of the disvalue of a bad outcome and the probability of its occurrence. And risk-benefit analysis, for the purposes of aligning expenditures to prevent costs with the benefits that prevention produces, requires a probability judgment of the occurrence of the cost.

There is a third kind of risk assessment not based on a priori knowledge, but on evidence and educated estimation. Exposure risk in insurance, for example, is often assessed when there is insufficient evidence about similar incidences or states of affairs to develop a statistically relevant probability of the event occurring. In these cases, knowledge gleaned from limited experience is combined with a theory about circumstances to develop an estimate of the probability of an outcome occurring. Insurers often rely on methods of this sort to insure large financial transactions. In all three of these cases, risks are a function, in part, of the probability of the outcome.

Some outcomes in life are uncertain in a more technical sense; namely, we are not in a position to reliably calculate their probability. This is the basis of the distinction between risk and uncertainty – a distinction, although crucial for thinking about future oriented action, not always well appreciated in policy discussions.[3] Our present understanding of the distinction is owed to the economist Frank H. Knight.

> The practical difference between the two categories, risk and uncertainty, is that in the former the distribution of the outcome in a group of instances is known (either through calculation a priori or from statistics of past experience), while in the case of uncertainty this is not true, the reason being in general that it is impossible to form a group of instances, because the situation dealt with is in a high degree unique.[4]

Knight suggests one source of uncertainty, namely, when we fail to have sufficient knowledge of a particular situation to reason by analogy to other instances that would establish a frequency-based probability of an outcome. In fact, however, there are other important sources of uncertainty as well.

If a person asks when he will have sufficient private savings to retire, the answer to this question is in at least one sense uncertain. It depends on how much he saves. Because the uncertainty is a function of human resolve and

---

3 This is a point driven home by Martin L. Weitzman in "A Review of *The Stern Review of the Economics of Climate Change*," *Journal of Economic Literature* XLV (2007): 703–724.

4 Frank H. Knight's, *Risk, Uncertainty, and Profit* (New York: Hart, Schaffner and Marx; Houghton Mifflin Co., 1921), chp. 8.

action, this can be called *moral uncertainty*. Various conditional scenarios can be devised. He can retire at age X, if he saves $A per month; but he can retire only at age X + n, if he saves only $A − m. When the uncertainty is regarding a single act, it disappears as soon as the person acts or loses the opportunity to do so. When it is about a set of actions, each independent of one another, uncertainty only vanishes once all of the actions have been taken. Demonstrable resolve to act supports ex ante confidence that the agent will act, and therefore mitigates uncertainty, but some uncertainty persists until the action is taken. This is often characteristic of diplomatic relations to end or forestall military conflict.

The person saving for retirement may confront another uncertainty as well. If market predictions are unreliable (given recent experience), then he may not be able to predict what the consequences of his choices will be. Lack of general knowledge of the workings of the market may undermine everybody's predictive powers. To distinguish this kind of uncertainty from the former, I refer to it as *epistemic uncertainty*. It is not ignorance about what our choices will be, but about their consequences. Epistemic uncertainty vanishes when the consequences of the action are known. Increased understanding of the workings of complex systems, such as financial markets, can transform uncertainty about the consequences of action into beliefs about the probability of various outcomes.[5] Insurers seek to transform epistemic uncertainty about individuals insured into beliefs about the probability of outcomes by grouping individuals according to various attributes that are relevant to risks.[6]

There are at least two different kinds of epistemic uncertainty. One kind derives from our contingent lack of knowledge of a process. Ignorance of this kind can, in principle, be overcome and the uncertainty transformed into risk. Another kind stems from the inability to predict outcomes in complex dynamic systems, which are chaotic. In such cases it is, in principle, impossible to know in advance what state of affairs will be produced at some future time.

Moral and epistemic uncertainties are independent of each other. One might have resolved to save $A, and rationally expect to be able to continue to do, but also have decided to put the money into volatile investments. In that case, there may be only epistemic uncertainty about how much savings will accrue and when (or if) he can retire. Or perhaps the savings is going into safe investments unaffected by financial turbulence. Then the uncertainty is transformed into risk, if there is some very low risk of bank collapse. In that case,

---

[5] In Knight's, *Risk*, chp. 8; increased power of prediction is the fourth of six ways discussed of dealing with uncertainty.

[6] *Ibid.* Knight has an extensive discussion of the method of consolidation.

there is some risk associated with his plan to retire at age X. If he has resolved to put his money into safe investments but can't be sure how much he will save, belief about the outcome is morally uncertain with respect to saving and probable with respect to the amount the savings will earn. When he will retire is, then, morally uncertain.

CLIMATE CHANGE AND UNCERTAINTIES

The predictions about the extent of the change to the climate system are beset with moral uncertainty because climate change is driven by concentrations of greenhouse gases in the atmosphere, and those concentrations depend on a myriad of policy choices about economic development and climate change mitigation. This is recognized by the Intergovernmental Panel on Climate Change (IPCC). In its *Fourth Assessment Report* (AR4), the IPCC considers various future scenarios or "storylines" regarding global economic development, integration, and technology use, each leading to different atmospheric concentrations of greenhouse gases.

The A1 storyline assumes a world of very rapid economic growth, a global population that peaks in mid-century and rapid introduction of new and more efficient technologies. A1 is divided into three groups that describe alternative directions of technological change: fossil intensive (A1FI), non-fossil energy resources (A1T) and a balance across all sources (A1B). B1 describes a convergent world, with the same global population as A1, but with more rapid changes in economic structures toward a service and information economy. B2 describes a world with intermediate population and economic growth, emphasising local solutions to economic, social, and environmental sustainability. A2 describes a very heterogeneous world with high population growth, slow economic development and slow technological change.[7]

Although climate policy analysts sometimes speak of a future business-as-usual baseline of unmitigated climate change against which mitigation policies are compared, the AR4 recognizes the fiction of such approaches. In fact, we are uncertain what the baseline will be in part because we do not know which macroeconomic policies affecting the direction of investment, growth, international trade, and population growth various countries will pursue.

These business-as-usual story lines have a range of likely warming from 1.1–2.9°C at the low end (B1) to 2.4–6.4°C at the high end (A1F1). I shall return to

---

[7] Intergovernmental Panel on Climate Change, *Climate Change 2007, Synthesis Report*, sec. 3.1. http://www.ipcc.ch/publications_and_data/ar4/syr/en/contents.html (accessed August 30, 2012).

the AR4's assumption of uncertainty with respect to the story lines presently. But assuming such moral uncertainty for a moment, the consequences for the moral assessment of climate change policy are significant. Moral uncertainty about which story line most accurately describes the future direction of the global economy produces uncertainty about temperature increase. When we seek to compare the moral advantages that climate change mitigation has over non-mitigation, the judgment about how bad the latter would be is subject to this uncertainty.

In the years since the AR4 was published, subsequent studies have suggested that the moral uncertainty about the direction of the global economy may be reducing. To some extent, the direction of global economic development is path dependent. Because of, among other things, long-term capital investments, development is more likely to continue along a path that it has been taking, unless by political resolution – to pursue more mitigation, for example – the path of development is altered. Several subsequent studies give reason to believe that the economy is moving along a high-emissions path, owing largely to the tremendous increase in fossil fuel use in China and other developing countries and insufficient CO_2 emissions reductions in the developed world.[8] The development of fossil fuel-based energy sources requires considerable capital investments. Absent an international agreement to mitigate climate change, countries will be loath to put to rest coal-fired plants before the investment in them has been recouped. So, there is good reason to believe that development along or near the A1F1story line will continue. This reduces considerably the range of uncertainty about temperature increase in a business-as-usual scenario, and it gives us good reason to assume that the status quo will lead to a temperature increase on the high end of the range.

The IPCC's Fifth Assessment Report (AR5) groups possible future states of the climate system according to four broad Representative Concentration Pathways.[9] Here, there is less focus on the moral uncertainty attached to the business-as-usual baseline. Rather, each pathway models an outcome that results from the positive radiative forcing of specific concentrations of greenhouse gases in the atmosphere. Positive radiative forcing is the increased radiant energy from the sun caused by greenhouse gases in the atmosphere. The IPCC counts one of the four scenarios as a mitigation scenario, two as stabilization scenarios, and only one as business-as-usual. With respect to the

---

[8] See also I. Allison, et al. *The Copenhagen Diagnosis* (Sydney: The University of New South Wales Climate Change Research Center, 2009), 9, fig. 1.

[9] Intergovernmental Panel on Climate Change, *Climate Change 2013, The Physical Science Basis, Summary for Policymakers*, pp. 17 and 27. http://www.climatechange2013.org/images/uploads/WG1AR5_SPM_brochure.pdf (accessed December 12, 2013).

latter, the mean projected temperature increase by end of the century is 3.7°C, within a range of 2.6–4.8°C.[10] That would amount to very rapid warming over the next eight decades. The uncertainty that attaches to that temperature range is epistemic, not moral, since it is uncertain how much warming to expect from a specific increase in radiative forcing.

Epistemic uncertainty is characteristic of both business-as-usual and mitigated climate change projections. Even if we set aside the moral uncertainty about how much we will emit and assume certain emissions levels, there is considerable uncertainty about the effects of the emissions. One fundamental source of this uncertainty concerns what climate scientists call *climate sensitivity*. Equilibrium climate sensitivity is expressed in degrees. It is the equilibrium temperature increase from preindustrial times of a doubling of CO_2 in the atmosphere. The AR5 maintains that climate sensitivity is likely to be in the range of 1.5 to 4.5°C.[11] The midpoint of this range is 3°C and the AR4 concluded that 3°C was the best estimate for climate sensitivity.[12] The source of the uncertainty is not whether CO_2 is a greenhouse gas; that has been understood since the nineteenth century. Rather, the climate system contains several feedback mechanisms, which amplify or counteract the warming caused by increased atmospheric CO_2, and the effects of many of these are not yet predictable. For example, an increase in temperature causes more water to evaporate. Water, when it forms insulating clouds in the lower atmosphere, is also a greenhouse gas. Hence, evaporation is usually thought of as a positive feedback mechanism, causing increased warming. But some accounts hold that cloud formation produces negative feedbacks, counteracting warming.

Predicting cloud activity is a massively complicated affair. According the AR4, "[c]loud feedbacks remain the largest source of uncertainty."[13] Indeed, given that global mean temperature increase and atmospheric CO_2 increase are well observed and documented, and the greenhouse properties of CO_2 have long been understood, the only hope of a scientifically respectable case for climate change skepticism would seem to be based on uncertainty regarding the feedback mechanisms, most especially that of increased cloud cover. One very controversial scenario has water evaporating in the tropics and forming rain clouds that drain most of their water in rainstorms. This leaves less water to form as ice crystals into higher level cirrus clouds. With fewer such clouds,

---

[10] *Ibid.*, 21.
[11] *Ibid.*, 14.
[12] IPCC, 2007 *Synthesis Report*, sec. 2.3. The AR4's range of likely climate sensitivity is slightly narrower, 2–4.5°C. It is good news that the AR5's range of climate sensitivity goes slightly lower to 1.5°C, all the more so since all the news about emissions growth is very bad.
[13] *Ibid.*

more long-wave energy would escape, countering other warming effects caused by evaporation. In the scientific literature, such an alleged negative feedback is referred to as *the iris effect*, but its existence is highly disputed.[14]

The consensus position is that water evaporation will function as a positive feedback mechanism. But cloud formation in response to global warming is, by all accounts, exceedingly difficult to predict. There are multiple competing accounts of the positive feedback mechanism without a single one having broad credibility. The impact of cloud formation on global climate change is an important instance of epistemic uncertainty.

Uncertainty about the precise measure of climate sensitivity within the likely range of 1.5–4.5°C is the reason for the uncertainty about how much the planet will warm with respect to a particular concentration of CO_2 in the atmosphere. Warming causes sea-level rise through thermal expansion and the melting of land-based ice sheets and glaciers that feed into the oceans. So, the uncertainty about warming is one factor in the uncertainty about the extent of sea-level rise. The AR5 estimates the range of sea-level rise for an end of the century warming of 2.2°C to be in the range of 0.33–0.63 m.[15] But that range does not express the full extent of uncertainty, which is exacerbated by a lack of understanding of dynamic ice-sheet collapse. Larsen B, an ice sheet in Antarctica about the size of the state of Rhode Island, had been stable for about 12,000 years before it collapsed during the course of several weeks in January and February 2002. There had been evidence that it was in trouble, but the speed of its demise was unexpected. As surface ice melts, water seeps into cracks and can lubricate the gaps between the ice sheet and land, creating a slide into the ocean. These events are not well understood and not predictable far in advance. Hence, the AR5 range of sea-level rise does not include "rapid dynamical change."[16]

In the summer of 2012, an ice sheet twice the size of Manhattan unexpectedly broke away from a glacier in northern Greenland. Dynamic land-based ice-sheet collapse has the potential of considerably extending the range of sea-level rise upward. Some scenarios warn of a possible sea-level rise in excess of 5 m

---

[14] An account of the iris effect hypothesis directed toward a nonspecialist audience can be found in Richard S. Lindzen, "Is the Global Warming Alarm Founded on Fact?" in Ernesto Zedillo, ed., *Global Warming: Looking Beyond Kyoto* (Washington, DC: The Brookings Institution Press, 2008), pp. 21–33. http://www-eaps.mit.edu/faculty/lindzen/L_R-Exchange.pdf (accessed September 6, 2012). There is little support in the scientific literature for the hypothesis. One critique contends that the evidence supports a weak, positive feedback from tropical cloud formation. See Bin Lin, et al., "The Iris Hypothesis: A Negative or Positive Cloud Feedback?" *Journal of Climate* 15 (2002): 3–7.

[15] IPCC, 2013 *The Physical Science Basis, Summary for Policymakers*, 21.

[16] *Ibid.*, fn. B.

this century.[17] To put that in perspective, 10 percent of the world's population lives at the altitude of not more than 10 m above sea level. This includes metropolitan centers such as New York City at 10 m, Miami at 2 m, Shanghai at 4 m, Dhaka at 4 m, Bangkok at 2 m, and much of Mumbai at 10–15 m.

The uncertainty about warming is more or less fixed to uncertainty within temperature ranges for a given concentration of CO_2 in the atmosphere. It is not that we have no good ideas about how much temperature increase there would be for a particular concentration of CO_2 in the atmosphere. The uncertainty about sea-level rise is somewhat less constrained because there are at least two factors at work: the uncertainty about temperature increase, which causes the thermal expansion of the oceans, and the uncertainty about the contribution of ice melting to the rise in sea level. It is worth stressing, however, that the latter uncertainty is about how much more sea-level rise there will be in addition to that caused by thermal expansion of the oceans. Even if we can fix a range of uncertainty for the first factor, there is the possibility that the rise could be considerably higher because of rapid ice sheet melting.

UNCERTAINTIES AND CATASTROPHES

Uncertainty about climate sensitivity is a significant factor in uncertainty about sea-level rise. Both of these are factors in additional uncertainties in the climate system, some of which present the possibility of major, and even catastrophic, irreversible change. This accumulated uncertainty is sometimes referred to as *cascading uncertainty*.[18] For example, the AR4 warns of massive species extinction (40–70 percent of those species assessed) if warming exceeds 3.5°C.[19] Positive feedbacks could produce a variety of climate-related catastrophes. Melting ice and thawing methane hydrate provide good examples of possible positive feedbacks that would make for major climate change. Melting sea ice affects the Earth's albedo, which is the measure of the amount of the Sun's radiative energy that is reflected back into space. As white sea ice melts into darker waters, the albedo becomes lower. The result is more surface absorption of radiative energy – a positive feedback.

[17] James Hansen, "Huge Sea Level Rises Are Coming – Unless We Act Now," *New Scientist* 2614, (July 25, 2007). http://www.newscientist.com/article/mg19526141.600-huge-sea-level-rises-are-coming-unless-we-act-now.html?full=true (accessed September 6, 2012).

[18] Stephen H. Schneider and Kristin Kuntz-Duriseti, "Uncertainty and Climate Change Policy," in Stephen Schneider et al., eds., *Climate Change Policy: A Survey* (Washington, DC: Island Press, 2002), pp. 67–68.

[19] IPCC, 2007 *Synthesis Report*, sec. 3.2.4.

The possible effects of land-based ice melting on the thermohaline circulation are alarming. The thermohaline circulation is a vast oceanic water transfer
system. As its name suggests, it is propelled by salt and heat. The thermohaline circulation includes the Gulf Stream in which warmer tropical water
moves north along the eastern seaboard of the United States. Warm water
continues eastward toward Europe as the North Atlantic Drift. In the North
Atlantic, incoming warmer water evaporates, which both cools and increases
the salinity of the remaining water. This cooler and heavier water sinks to the
ocean floor and heads south. It eventually upwells again either in the Southern
Ocean or the North Atlantic. The North Atlantic Drift is a major factor in the
warming of Northern Europe that produces a more temperate climate than
corresponding latitudes in North America. For example, Toronto is about the
same latitude as Marseille. The latter owes its temperate climate to the North
Atlantic Drift.

The massive Greenland ice sheet is second only to Antarctica in size. It
melts each summer and then rebuilds each winter. During the summer of
2012, there was more observable melting across its surface than at any time
since satellite photos began recording it thirty years ago.

The melting of the Greenland ice sheet threatens to raise sea levels by
as much as 6.5 m.[20] But it also poses a threat to the Atlantic Drift and the
temperate climates of Western Europe. The introduction of unprecedented
amounts of fresh water into the North Atlantic would diminish water salinity
and slow, or even halt, the conveyance of water to deep depths that draws the
warm water from the south. Recognition of the possibility of the shutdown of
the Atlantic Drift is not new. Scientists have generally estimated the probability
of it happening to be very low. The AR4 holds it to be very unlikely that a
complete shutdown will occur, but very likely that circulation will slow.[21]
However, if the Greenland ice sheet looses considerable ice coverage, we will
be in an unprecedented situation with uncertain effects on the Atlantic Drift.
Weather in Western Europe would be more akin to Canada. This would not
only be a major inconvenience, it would dramatically increase energy use
caused by increased winter heating. And increased fossil use would constitute
a positive feedback.

Methane is a greenhouse gas that is 25 percent more effective than CO_2
at trapping the Sun's radiative energy. About 60 percent of the methane in
the atmosphere is the result of a variety of human activities, such as livestock

---

[20] Richard Z. Poore, et al., "Sea Level and Climate," United States Geological Survey Fact Sheet
    020–00. http://pubs.usgs.gov/fs/fs2--00/ (accessed July 4, 2013).
[21] IPCC, 2007 *Synthesis Report*, sec. 3.2.4.

ranching and coal mining. There are also natural sources of atmospheric methane in wetlands and decomposing forests, and vast stores of it in the Arctic tundra and in the form of methane hydrate crystals in the floor sediment of the Arctic Ocean at great depths. According to analyses of air bubble samples drawn from ice sheets, concentrations of methane in the Earth's atmosphere are higher now than at any time during the past 400,000 years. Concentrations have more than doubled over the last two centuries. Warming land and water is causing additional methane to be released – a further positive feedback. Since the 2007 publication of the AR4, hundreds of plumes of methane have been observed rising from the Arctic Ocean.[22] Because ocean warming lags behind atmospheric warming, this release of methane may continue even if the atmosphere were to stabilize at the current temperature. And should these releases become more widespread, they could increase the amount of methane currently released globally from natural sources by about 10 percent.

Both methane and CO_2 are stored in the frozen northern tundra. More than 150,000 seeps of methane from thawing tundra have been observed bubbling up from lakes in Alaska and Greenland. So far, Canada and Russia have not been observed, so the total amount seeping into the atmosphere is not known.[23] One recent geologic study of ancient ocean sediment suggests that a methane release from ice sheets 635 million years ago produced sufficient warming then to release oceanic methane hydrate, which resulted in an abrupt warming of the planet, bringing to an end the period in which the Earth suffered one of its most severe ice ages.[24] The lead author of this study, geologist Martin Kennedy, compares that process to current global warming:

> We are witnessing an unprecedented rate of warming, with little or no knowl-
> edge of what instabilities lurk in the climate system and how they can influ-
> ence life on Earth. But much the same experiment has already been con-
> ducted 635 million years ago, and the outcome is preserved in the geologic
> record. We see that strong forcing on the climate, not unlike the current
> carbon dioxide forcing, results in the activation of latent controls in the

[22] National Oceanography Centre, Southampton, "Warming Ocean Contributes to Global Warming," August 14, 2009. http://www.noc.soton.ac.uk/nocs/news.php?action=display_news&idx=628 (accessed August 31, 2012).

[23] National Snow and Ice Data Center, *Icelights: Your Burning Questions about Ice & Climate*, "What Does Seeping Methane Mean for the Thawing Arctic?" July 3, 2012. http://nsidc.org/icelights/2012/07/03/what-does-seeping-methane-mean-for-the-thawing-arctic/ (access August 31, 2012).

[24] Martin Kennedy, et al., "Snowball Earth Termination by Destabilization of Equatorial Permafrost Methane Clathrate," *Nature* 453 (2008): 642–645.

climate system that, once initiated, change the climate to a wholly different state.[25]

We do not know how much methane is likely to be released as the planet warms because we do not know how much it will warm or what the precise relationship between warming and methane release is. But the theory suggested by this study of prehistoric climate change suggests grave dangers.

DANGER AND UNCERTAINTY

In the previous two sections we have seen that there is a range of uncertainty about the expected effects of climate change. Some of this is because of epistemic uncertainty about the precise value of climate sensitivity. The equilibrium mean surface temperature of the Earth for any given concentration of CO_2 in the atmosphere is uncertain within a range of degrees. Factors such as cloud formation and changes to the Earth's albedo (caused by ice melting) create a range of uncertainty about how much warming to expect for any particular business-as-usual scenario. Policy makers also must assume a particular business-as-usual scenario and there is some moral uncertainty about that. But there is significant evidence to suggest that absent additional climate change mitigation, we are on a path of high fossil fuel consumption, which could produce warming up to 6.4°C.

In Chapter 1, the discussion of the personal analogy established that prudential judgments about danger are normative. The judgment that a course of action is dangerous, or too risky, rests on a reason why, in light of what a person has reason to value, the action should not be pursued. Judgments about danger in the case of climate change are not prudential, but moral, because the lives and well-being of billions of people are affected. In that chapter we identified three categories of reasons relevant to considering whether a fossil fuel-intensive energy policy with higher risks of climate change was dangerous.

1. Reasons for mitigation: The reasons that future people would have to avoid the risks of climate perturbations.
2. Reasons for fossil fuel consumption: The reasons that current people have to consume energy, especially to eradicate poverty.

[25] "Large Methane Release Could Cause Abrupt Climate Change As Happened 635 Million Years Ago," *Science Daily*, May 29, 2008. http://www.sciencedaily.com/releases/2008/05/080528140255.htm (accessed August 31, 2012).

3. Reasons for fossil fuel consumption: The reasons that future people would have for poverty-eradicating human development to commence, continue, or accelerate.

We can now appreciate a simplification that was built into the discussion in Chapter 1. The concepts of risk and uncertainty were not distinguished. In fact, included under the category of the risks of climate change were both risks and uncertainties. I argued that climate change policy was dangerous, in the sense that it should be avoided, if its costs made the prospects for poverty eradication, now and in the future, worse than other available policy alternatives. But prospects may be either probabilities or uncertainties.

The account of dangerous climate change in Chapter 1 is consistent only in part with a risk-benefit approach. And in any case, no such approach can work where no probabilities can be attached to an outcome. This is the case with uncertainties in the technical sense. As we have seen, we confront several uncertainties when it comes to forecasting the effects of climate change.

We have already observed warming of 0.6–0.9°C (1.1–1.6°F) from 1906 to 2005, and, according to the National Aeronautics and Space Administration (NASA), "the *rate* of temperature increase has nearly doubled in the last 50 years."[26] This observed warming is not the whole story. Climate scientists refer to a "warming commitment" as the amount of additional warming that we are already locked into at present concentrations of CO_2 in the atmosphere. Most of the Earth's surface is covered with water, and water warms more slowly than air. If greenhouse gases contributed nothing more to the warming of the planet, our warming commitment by the end of this century would be an *additional* 0.4–0.6°C (0.72–1.08°F). Taking both ranges into account with the present level of CO_2 in the atmosphere, we are committed to an eventual warming of between 1 and 1.5°C. By 2100, this would produce a sea-level rise of at least 11 cm (4.3 in.) from thermo-expansion alone.[27] A rise of 10 cm (3.93 in.) per century would be expected also for centuries to come.[28] All of this is based on the currently unrealistic assumption that concentrations of CO_2 will not increase. Not only were CO_2 concentrations increasing in the atmosphere for four out the five decades since 1960 (the exception being the 1990s), the increase has

[26] NASA, Earth Observatory, "Global Warming." http://earthobservatory.nasa.gov/Features/GlobalWarming/page2.php (accessed September 5, 2012).
[27] Gerald A. Meehl, et al., "How Much More Global Warming and Sea Level Rise?" *Science* 307 (2005): 1769–1772.
[28] T. M. L. Wigely, "The Climate Change Commitment" *Science* 307 (2005): 1766–1769.

been occurring at a faster rate. It has doubled in the last fifty years.[29] Imagine a car accelerating, with the increase in the rate of acceleration also becoming greater. Then imagine that it is headed for a cliff.

Much of what I have summarized is simply known from very careful scientific observation of the temperature increase to date, sea-level rise to date, CO_2 concentration increases to date, and increases in the rate of CO_2 concentration growth. None of this incorporates any forecasting uncertainty; we are uncertain about neither warming nor sea-level rise to date. However, we are uncertain about how much more to expect regardless of whether or not CO_2 concentrations increase. One way that climate scientists attempt to arrive at greater confidence in their future projections is to use models, including various values of climate sensitivity and a host of other variables, retrospectively, to see how well they track observed changes over the course of the twentieth century. The models used for future projections are those that correspond well to observed changes over the course of the twentieth century.[30] So, when climate scientists use models to forecast that unmitigated climate change is likely to give rise to a global mean-temperature increase from 1.1°C on the low end of the best-case business-as-usual scenario to 6.4°C on the high end of the worst-case business-as-usual scenario (which there is evidence that we are following), confidence in this range is strengthened because the models have corresponded well to already observed warming.

UNCERTAINTY, CLIMATE SENSITIVITY, AND POLICY

One troublesome feature of international calls to mitigate climate change by pursuing the goal of limiting warming to 2°C (3.6°F) is that because of uncertainty about climate sensitivity, we do not know what the maximum concentration of CO_2 in the atmosphere should be to maintain that limit. Recall that the IPCC put the best estimates of climate sensitivity at 3°C (5.4°F), but within a possible range of 1.5–4.5°C (3.6–8.1°F). Using 3°C as the measure of climate sensitivity, atmospheric concentrations of CO_2 must peak within the range of 350–400 parts per million (ppm) in order to keep warming near 2°C (possibly going as high as to 2.4°C).[31] As of the writing of this book, concentrations have risen to more than 400 ppm. In other words,

---

[29] National Oceanic and Atmospheric Administration (NOAA) Earth Systems Research Laboratory, Global Monitoring Division, "Trends in Carbon Dioxide." http://www.esrl.noaa.gov/gmd/ccgg/trends/mlo.html#mlo_full (accessed September 5, 2012).

[30] See figure SPM 4 in IPCC, 2007 *Synthesis Report*, sec. 2.

[31] *Ibid.* sec. 5.4.

we are already at the threshold of that range that gives a decent chance of limiting warming to 2°C.[32] Because CO_2 stays in the atmosphere for so long, cumulative historical emissions are the chief factor in the concentration of CO_2 in the atmosphere.[33] Arresting the temperature increase requires not only reducing global emissions, but also an eventual no-carbon economy.

The concentrations of CO_2 in the atmosphere that correspond to a 2°C global mean temperature increase depend on the correct measure of climate sensitivity. If it is 2°C (3.6°F) near the low end of the consensus range, then atmospheric concentrations could increase to 558 ppm before leveling (because this 558 ppm amounts to a doubling of CO_2 and – as you may recall – climate sensitivity is a measure of the temperature increase at such a doubling). That would be consistent with a 10–60 percent increase in emissions against 2000 levels.[34] As a result of emissions increases since 2000, emissions in 2011 were near the middle of that range.[35] Hence, if climate change is near the low end of the consensus range, we must simply find a way to stop increasing emissions rather than decreasing them. But if that were our policy, and if climate sensitivity in fact is 3°C, then hitting the 558 ppm target would lead to a temperature increase of 3.2–4°C (5.76–7.2°F). There are also claims about climate sensitivity outside of the range of the existing scientific consensus; one goes as low as 0.5°C (0.9°F).[36] If we took this as correct, there would be no point in seeking any emissions reductions. If we were to base our policy on such an account, and if climate sensitivity is actually 3°C, then our policy would probably lead to consequences as predicted by the various business-as-usual scenarios, which include catastrophic temperature increases of up to 6.4°C.

Policy makers cannot hope to settle the science and be rid of the matter of epistemic uncertainty. So, what is the morally appropriate policy response to this uncertainty? There are two mutually reinforcing reasons why policy should be based on the middle-range consensus position of 3°C. The first is that it is the position represented by the scientific consensus. The second is that the costs of being wrong about this, if policy is based on a lower measure of climate sensitivity, are intolerably high. Let's consider both reasons.

---

[32] National Oceanic and Atmospheric Administration (NOAA) Earth Systems Research Laboratory, Global Monitoring Division, "Trends in Atmospheric Carbon Dioxide." http://www.esrl .noaa.gov/gmd/ccgg/trends/ (accessed September 10, 2012).

[33] IPCC, 2013 *The Physical Science Basis, Summary for Policymakers*, 25.

[34] IPCC, 2007 *Synthesis Report*, sec. 5.4.

[35] Using Emission: The European Commission's Data Base For Global Atmospheric Research (EDGAR). http://edgar.jrc.ec.europa.eu/overview.php?v=CO2ts1990-2011 (accessed September 10, 2012).

[36] See Lindzen, "Global Warming: Looking Beyond Kyoto."

The first reason to base policy on the middle-range consensus position on climate sensitivity is that policy makers are not qualified to make decisions about matters of scientific disagreement and uncertainty. Their role is not to attempt to settle the science. In some ways, the position of a policy maker is not different in this regard than a patient deliberating in light of reading conflicting accounts about the safety of a course of medical treatment. If one wonders whether to pursue a particular therapy, the most rational course for the vast majority of medical patients is to simply trust the consensus position of doctors and medical researchers regarding the safety of the medicine.[37] It may be important to study the nontechnical details of the matter, not with the hope of settling any scientific uncertainty regarding the medicine, but simply to understand whether there is any controversy and if any of the factors apply to one's medical condition. But it would be remarkably imprudent for the vast majority of patients to suppose that they know better than the medical community. The only rational position in most cases is to trust the authority of the relatively settled scientific position.

There is at least one important way in which the situation of the policy maker is different than that of the patient confronting a choice about her medications. In the choice the patient makes, norms of prudence apply. But the choice that policy makers confront about which measure of climate sensitivity to accept affects the lives of billions of people. If they fail to base their policy decision on the position of the scientific consensus, they are similar to a guardian who makes a decision about medical care for the person to whom he is entrusted in disregard of the medical consensus. The criticism that the guardian is subject to in such a case is not that he was imprudent, but rather that he was morally irresponsible.

The appropriate attitude of the patient and the policy makers in these cases is not really a belief in the truth of the scientific views, but one of trust in those who are in a much better epistemic position and who mostly agree on the appropriateness of a course of action. An attitude of trust in the scientific community is indirectly related to the truth of the position on which one is acting. One trusts that they are in a better position to detect falsehoods and to get nearer the truth. There can, of course, be times when such trust is inappropriate. One possibility is that one has very good reason to believe that there is systematic bias that one is somehow privy to understanding, and which escapes the understanding of the scientists involved. But it is worth stressing what an unusual position that would be. It would have to be the case

---

[37] On this point, see also Mike Hulme, *Why We Disagree about Climate Change* (Cambridge: Cambridge University Press, 2009), p. 86.

that one understood enough of the science and enough of the workings of consensus formation in the profession to have good grounds for distrusting the consensus. In most cases for most nonspecialists, the evidence would simply be unavailable. Alternatively, if there was not really a scientific consensus on the matter, but a wide range of opinions, one would simply be wrong to assume that there was a consensus.

Notice that I am not arguing that scientists who dissent from the consensus are wrong in pressing their views. On the contrary, science typically proceeds by scientists coming up with convincing evidence in support of theories that challenge the accepted view. In many cases, scientists can be viewed as serving the cause of greater understanding by pressing against the accepted wisdom. Most of them, however, will be wrong. But, of course, with the long view of the history of science in mind, most scientific consensus is wrong as well. It is noteworthy that the scientist defending an unconventional view is not forced by epistemic limitations simply to trust the consensus view, as are most of us when it comes to most matters of scientific consensus.

The second reason that policy makers should accept the consensus position on climate sensitivity is based on the dire costs of being wrong. If policy makers were to gamble that the low end of the scientific consensus on climate sensitivity is correct or that something even lower is correct, and if that turns out not to be the case, they would be exposing billions of people to significant, although indeterminate, probabilities of grave outcomes. One important consideration is the unprecedented rate at which temperature increase would occur if policies were based on a climate sensitivity value of 2°C (3.6°F) or less, and if climate sensitivity in fact is 3°C (5.4°F). "When global warming has happened at various times in the past two million years, it has taken the planet about 5,000 years to warm 5 degrees. The predicted rate of warming for the next century is at least 20 times faster."[38] Such warming would put unprecedented stress on ecosystems, agriculture, and human communities.

One comprehensive study of the effects on agriculture of a 5°C (9°F) warming projects a loss of global agricultural yields of 3–16 percent by 2080, depending on whether the increased CO_2 in the atmosphere helps to fertilize a sufficient number of crop types.[39] According to this forecast, industrial countries would suffer the least. Without the beneficial effects of CO_2 fertilization – which are uncertain – crop losses in these countries would be about 6 percent;

[38] NASA, Earth Observatory, "How Is Today's Warming Different from the Past?" http://earthobservatory.nasa.gov/Features/GlobalWarming/page3.php (accessed September 6, 2012).
[39] See William Cline, *Global Warming and Agriculture* (Washington, DC: Peterson Institute, 2007). A summary of the study's results is available at: http://www.imf.org/external/pubs/ft/fandd/2008/03/pdf/cline.pdf (accessed September 6, 2012).

with the benefit that yields could improve by 8 percent. Africa would suffer the most with losses ranging from 17 to 28 percent. These global crop-yield losses are especially alarming in the context of a growing global population. With mid-century forecasts of a global population of 9 billion people, the population is expected to be 29 percent larger than it presently is. The combination of population growth and globally rising income is expected to triple food demand by 2080. Coping with the countervailing forces of decreasing crop yields and increasing food demand in such a short time frame might be extremely difficult. As is the case for many other climate change threats, geography and the economy conspire to make the threats especially great to the global poor.

The effects of such a rapid warming on ecosystems would be devastating. Numerous species would fail to adapt. Recall that with a temperature increase of 3.5°C (6.3°F), mass extinction – possibly exceeding two-thirds of known species – is projected.

We could pick out several threats of global warming in the range of 4–5°C (7.2–9°F), but to simplify the discussion, I'll focus on the threats of widespread food insecurity and massive species extinction. If we continue on a high emissions business-as-usual path, and if the best estimation of the consensus of scientific community is correct, there are significant probabilities of severe food stress and mass extinction with such warming. If the few scientists engaging in climate research who reject the consensus view are right, then efforts to mitigate climate change and save for adaptation will be a waste of money. The opportunity costs of that might include less money available to devote to financing human development. But this is not necessarily the case for two reasons. First, the distribution of costs is itself a matter of policy. The costs of climate change–mitigation policies can be distributed in a manner that allows developing and least-developed states to pursue macroeconomic policies with human development objectives. Second, adaptation savings generally can be used for projects that are consistent with human development. In fact, in developing and least-developed countries, climate change adaptation will largely be about protecting developmental gains and seeking to advance developmental objectives in any case. So, there is no fundamental tension between pursuing climate change adaptation and human development.

Estimates of the total costs of climate change mitigation vary considerably. In part, this variance is owing to the moral uncertainty of the status quo and the epistemic uncertainty of the impact of climate change. But it is also owing to controversies surrounding how much to discount the future costs of climate change. This matter is taken up in Chapter 4. For now let's consider the costs estimated in the *Stern Review: The Economics of Climate Change* (the

Stern Review) commissioned by the British Government.[40] The *Stern Review*, which advocates extensive and immediate action to mitigate climate change, puts these costs at 1 percent per year of the global gross domestic product.[41] In a separate study, Nicholas Stern argues that the costs may be higher, around 2 percent of the global domestic product.[42] Two aspects of this estimation of the costs of mitigation are noteworthy. First, it is higher than in many studies both because Stern advocates extensive mitigation and because the social discount rate that he employs takes the costs of unmitigated climate change to be high.[43] So, invoking Stern is not to endorse his view, but to give an example of a fairly high-cost assessment of the appropriate mitigation policies. Second, the cumulative economic effects of even such a high-cost assessment are slight. "[T]he world economy would take roughly an additional six months in reaching the level of world income it would otherwise reach by 2050 (assuming high-carbon growth were sustainable for that long)."[44]

The combination of high stakes and the relatively low costs of avoiding the worst outcome encourage focusing on the worst-case scenarios of choosing between the mid-range consensus view of climate sensitivity and views either at the low end of the consensus view or below it. If the mid-range consensus among climate scientists is correct, the moral costs of not mitigating could be very high: widespread food insecurity – especially among the poor – and massive species extinction. The cost of mitigation is a minor loss of global GDP. Alternatively, if climate sensitivity is really very low, the moral cost of climate change will be low, and the money spent on mitigation based on the trust that the scientific consensus about 3°C is correct will have been wasted. The costs of erring on the side of trusting in the higher measure of climate sensitivity are considerably lower than the costs of erring on the side presuming a lower value. And that is relevant to morally responsible policy making.

The considerations of trust in the scientific consensus and the terrible costs of being wrong converge to support basing policy on a climate sensitivity value of 3°C. These considerations, however, could pull apart.[45] For

---

[40] Nicolas Stern, *Stern Review: The Economics of Climate Change*, online edition. http://www.hm-treasury.gov.uk/sternreview_index.htm (accessed October 5, 2012).

[41] *The Stern Review of the Economics of Climate Change*, Summary of the Conclusions, vii. http://www.hm-treasury.gov.uk/d/Summary_of_Conclusions.pdf (accessed September 6, 2012).

[42] Nicholas Stern, The Global Deal: Climate Change and the Creation of a New Era of Progress and Prosperity (New York: Public Affairs, 2009), p. 52.

[43] For a rival account of the costs of mitigation, which takes the costs of the optimal policy to be lower, see William Nordhaus, *A Question of Balance: Weighing the Options of Global Warming Policies* (New Haven, CT: Yale University Press, 2008).

[44] *Ibid.*, 54.

[45] I am indebted to Tim Diette for pressing this point.

example, it would be even more cautious to base policy on a climate sensitivity value of 4.5°C, on the high end of the consensus range. This would require, however, much more significant climate change mitigation to keep warming at or below 2°C, raising the costs of such a policy considerably. Recognition of this highlights an additional reason, mentioned in the preceding discussion, for basing climate policy on a climate sensitivity value of 3°C; namely, that in comparison to the possible costs of being wrong, the costs of acting on this policy are very modest. These two considerations are important to the justification of acting in accordance with a strong precautionary principle.

STRONG PRECAUTION

As I noted at the beginning of this chapter, the UNFCCC states the following principle: "Where there are threats of serious or irreversible damage, lack of full scientific certainty should not be used as a reason for postponing such measures." There are many formulations of a commitment to precaution; the UNFCCC's principle states a relatively weak version of the precautionary principle. Application of this principle takes note of the high stakes of not pursuing a policy of mitigation based on a value of climate sensitivity of 3°C and the relativity low costs of doing so. Invoking the principle would support the view that uncertainty about whether widespread food insecurity and massive species loss will occur, if policy is pursued on the assumption of a lower-value climate sensitivity, is not a reason for not pursuing policy based on the 3°C value.

The argument that I made in the previous section, however, suggests a stronger version of precaution. I have been arguing not merely that uncertainty provides no reason not to pursue a policy of mitigation, but that it provides an important reason to pursue such a policy. This version of the precautionary principle is an instance of reasoning according to the minimax rule – a rule of thumb for rational choice under certain kinds of uncertainty on the assumption that an agent is seeking to minimize her loss. The rule holds that between courses of action – all with uncertain negative outcomes – the agent should compare only the highest loss scenarios of the courses and choose the course of action that causes the lowest of the highest loss scenarios to come to pass.[46]

---

[46] On minimax, see also William Fellner, *Probability and Profit* (Homewood: Richard D. Irwin, Inc., 1965), pp. 140–142. I emphasize that the path causes the payouts in order to avoid complications that occur when thinking about payouts that are not caused by one's decisions. See Robert Nozick, "Newcomb's Problem and Two Principles of Choice" in Nicholas Rescher, ed., *Essays in Honor of Karl G. Hempel* (Dordrecht: Kluwer, 1969), pp. 114–146.

TABLE 3.1. *Possible net payouts (possible benefits minus possible costs)*

	Possible net payout 1	Possible net payout 2	Possible net payout 3
Course One	5	−2	−12
Course Two	−10	−3	1

In other words, the rule recommends acting so as to minimize the maximum loss. This is illustrated by considering Table 3.1.

Table 3.1 shows three possible net payouts caused by two different courses of action. The minimax rule recommends comparing these two courses of actions according to their highest possible (but uncertain) negative payouts – in this case, Net Payout 3 of Course One compared to Net Payout 1 of Course Two – and choosing the course with the lowest–highest negative payout. Therefore, minimax recommends Course Two. This is a kind of precautionary recommendation. It is not merely that uncertainty provides no reason *not* to act; rather, it provides a reason *to* act.

It would be implausible to claim that under conditions of uncertainty, rational agents should always seek the path whose consequence is the lowest maximum negative payout. The numbers in Table 3.1 could be read merely as an ordinal ranking of the net payouts, according to an agent's rational evaluation of the outcomes. But a mere ordinal ranking tells us nothing about how much reason the agent has to care about not receiving Net Payout 3 of Course One. And without knowing that, we don't know whether the agent has sufficient reason to forego the opportunity of Net Payout 1 of Course One because of the possibility of Net Payout 3. If the agent has very good reasons to care a great deal about getting Net Payout 1 and little reason to care about not getting Net Payout 3, then Course One could be worth the uncertain gamble. Alternatively, if the possible payouts are gains and losses in pennies, perhaps the rational choice is to pursue the one with the highest average net payout. This would amount to treating each of the three outcomes on each course of action as equally likely. Again, the choice would be Course One. As long the stakes are low and there is a continuum of losses, it might be sensible to treat the three possible outcomes for each course of action as equally probable. The moral of the story is that a mere ordinal schedule of the ranking of uncertain outcomes alone cannot justify the use of minimax.[47]

---

[47] See also Cass R. Sunstein, *Laws of Fear* (Cambridge: Cambridge University Press, 2005), p. 110 and Fellner, *Probability*, 142.

There are, however, very good reasons to follow the minimax rule under the right conditions. Suppose a twisted experimental philosopher arranges a study that assigns participants the following set of possibilities, all with unknown probabilities attached for each of the two courses of action in Table 3.1: All of the net payouts except Net Payout 3 of Course One represent net gains and losses in chocolate chip cookies whereas Net Payout 3 of Course One represents a fifteen-second shock with a cattle prod. No one would seriously deny the rationality of the approach recommended by the minimax rule in that circumstance. To generalize from that example, two conditions stand out: 1) An agent must have very good reasons to avoid the lowest net payout from among the available courses of action, and 2) in comparison to the reasons for avoiding the lowest net payout, the reasons for increasing the payout above the lowest are not nearly as compelling. These may be necessary for reasonable employment of the minimax rule, but perhaps they are not sufficient.[48]

No agent has reason to worry about a merely logical possibility. A threshold of probability, even if not strictly quantifiable, is necessary. At the least, the possible outcome should be explicable, even if not predictable, using standard scientific accounts. In any circumstance of uncertainty, a plausible case for the minimax approach would seem to require a third condition: 3) In order to be taken seriously as a real – even if epistemically uncertain – possibility, there must be a plausible explanation of its occurrence. In the previous example, the plausible explanation of each of the three outcomes for the two courses of action is that the twisted experimental philosopher has (credibly) announced that he will bring about one of the outcomes.

When the ice sheet Larsen B collapsed, no one thought it was miraculous. In retrospect, the collapse is largely explicable. Although scientists are prevented from making predictions about rapid ice sheet collapse, the uncertainty is caused neither by a lack of conceptual tools nor by ignorance of the basic physical laws at play. The complexity of the factors and the lack of ability to make the needed observations currently frustrate the capacity of scientists

---

[48] John Rawls notices this in his discussion of the corresponding maximin rule. See John Rawls, *Justice as Fairness: A Restatement* (Cambridge, MA: Harvard University Press, 2001), p. 98. Minimax and maximin are principles of the same form, with minimax directing action with respect to possible losses – minimize the maximum loss – and maximin directing action with respect to possible gains – maximize the minimum. (For Rawls, it is the gains of social cooperation that are at issue.) He notices that maximin, under uncertainty, would make sense only if (a) persons had no reason to care much about gains above the best minimum (the guaranteed level) and (b) the difference between the best minimum and others was significant. A useful application of Rawls's argument to the precautionary principle can be found in Stephen M. Gardiner, "A Core Precautionary Principle," *Journal of Political Philosophy* 14 (2006): 33–60.

to make medium-term predictions.[49] In addition to the minimum threshold of explicability, we have even more reason to be attentive to the possible outcome, if circumstances are producing some of the known causal factors.[50] The support for following a minimax approach with respect to climate change mitigation, then, is supported not merely by how extremely bad the possible worst outcome is and our relative indifference toward the better ones, but also by our reasonable beliefs that warming is causing such outcomes to be more likely. Such a belief can be perfectly reasonable without knowing with any exactitude how much more likely the outcomes are being made or where the threshold is.

We take the threat of food stress and massive species extinction seriously because we can reasonably foresee that continued warming will disrupt growing patterns for many crops, and we know that species are already migrating in response to warming. Widespread food stress and massive species loss are the terrible possible outcomes of a policy that fails to mitigate climate change adequately because it is based on a value for climate sensitivity that is too low. Compare that set of outcomes to the following: Pursuit of unnecessary mitigation, based on an assumed climate sensitivity value of 3°C, in an effort to limit warming to 2°C. Such mitigation would decrease global GDP by 1–2 percent, but would be unnecessary, at least in part, if climate sensitivity is in fact lower than 3°C. Because the first set of outcomes is so much worse than the second, it is perfectly reasonable to avoid the worst-case scenario as the minimax rule advises.

Someone might object that I have set a false dilemma by assuming that the choice is between, on the one hand, a precautionary approach to the climate policy based on a value for climate sensitivity of 3°C and, on the other hand, a policy based on a value of climate sensitivity of 2°C or lower. Perhaps an alternative is simply to wait and see if our knowledge of climate sensitivity becomes more precise. If we wait, and let observations and research continue, we will eventually obtain a richer understanding of the various feedback mechanisms that currently block our attempts to understand the likely temperature increase of a doubling of CO_2 in the atmosphere. This would allow us the option of mitigating later, while not incurring an unnecessary cost now if mitigation is

---

[49] See also Schneider and Kuntz-Duriseti, "Uncertainty and Climate Change Policy."

[50] These two points about knowing what the causal factors are and how or when they are coming to pass are also stressed by Henry Shue in his second of three features sufficient for ignoring probabilities or the lack thereof in "Deadly Delays, Saving Opportunities: Creating a More Dangerous World?" in Stephen Gardiner, et al., *Climate Ethics* (Oxford: Oxford University Press, 2010), p. 148.

not needed.[51] In the face of such uncertainty, perhaps the best policy approach is to await greater understanding. It is often a reasonable response to epistemic uncertainty to seek further knowledge in an effort to transform uncertainty into a risk assessment. This objection is not, however, what it might appear to be at first blush. It is not a third alternative. Instead, it amounts in practice to embracing the horn of the dilemma that argues for basing policy on a lower value of climate sensitivity because the practical implication of doing that would be roughly the same as waiting. Either way, policy would lead to a failure to pursue serious mitigation. And the possible consequences of doing that if climate sensitivity is 3°C are alarming. It is no rejoinder to my response to say that mitigation can be ramped up later if climate sensitivity turns out to be 3°C or higher. In the meantime, concentrations of CO_2 in the atmosphere would have continued to grow far beyond 400 ppm. This would drastically increase the technological difficulties and economic costs of maintaining the 2°C warming limit.[52] For, in the meantime, in the absence of a mitigation policy that raises the costs of greenhouse gas emissions, there would be fewer incentives for technological changes that would eventually make it cheaper to reduce emissions.[53]

OBJECTIONS AND RESPONSES

I have argued that uncertainty can be a reason for taking precautionary action if the following three conditions are met: 1) An agent has very good reasons to avoid the lowest net payout from among the available courses of action; 2) in comparison to the reasons for avoiding the lowest net payout, the agent's reasons for increasing the payout above the lowest are not nearly as compelling; and 3) there is a plausible explanation of the occurrence of the worst possible outcome. The precautionary principle has been the subject of much criticism. Often the criticism is directed toward stronger versions of the principle than that stated in UNFCCC. But I have been defending a stronger version, based on the minimax rule, which has us compare the uncertain outcome of

[51] Seeking to preserve option value as a response to uncertainty is defended in Christian Gollier and Nicolas Treich, "Decision-Making under Scientific Uncertainty: The Economics of the Precautionary Principle," *Journal of Risk and Uncertainty* 27 (2003): 77–103.

[52] See also Carl F. Cranor, *Regulating Toxic Substances: A Philosophy of the Science and the Law* (New York: Oxford University Press, 1993), pp. 25–28 for a general discussion of some of the risks of a status-quo bias in postponing regulation until scientific uncertainty is overcome.

[53] See also Richard A Posner, *Catastrophe: Risk and Response* (Oxford: Oxford University Press, 2004), p. 161.

food stress and species extinction, as representatives of many other very bad outcomes, against a possibly unnecessary modest decrease in global GDP if mitigation is pursued but not needed. Given these two possibilities, following the minimax rule is reasonable. We have reason to care very much about avoiding the highest maximum cost of two courses of action and little reason to care much about the costs that are less than that. Following the minimax rule is all the more reasonable because there is good evidence that the causal antecedents exist that could produce the highest maximum cost if we fail to take mitigation seriously. Hence, we are justified in taking precaution to mitigate climate change on the assumption that climate sensitivity is 3°C.

I now consider three prominent critiques of the precautionary principle to assess whether the problems attributed to it afflict the account that I have offered. Cass Sunstein criticizes certain strong formulations of the precautionary principle as paralyzing. As an example, he considers the principle that "regulation is required whenever there is a possible risk to health, safety, or the environment, even if the supporting evidence remains speculative and even if the economic costs of regulation would be high."[54] The problem with such a principle, according to Sunstein, is that regulation may also pose risks of the kind mentioned in the principle.[55] The more we increase safety standards for medicines and food, the less we have available for the sick and hungry. There is often no way to act that does not raise the possibility of harms of these sorts.

The defense of the minimax rule that I offered earlier, however, applies only to a much smaller set of circumstances. These are characterized by three features: We have good reasons to want to avoid the worst outcome; we care much less about the costs that are less than the worst; and one, and only one, course of action is known to produce some of the causal antecedents of the worst outcome. There is nothing paralyzing about a principle that would have us avoid the worst outcome in that case because the reasons to reject the one course of action do not apply to the others.

Bjørn Lomborg argues that the precautionary principle fails to provide us with any new advice because any sensible policy choice will have to "face the trade-offs between the costs of avoiding future harm and the cost of incurring harm right now. We are back then to discussing the cost-benefit trade off."[56] Although it is the case that the minimax rule requires circumstances in which we foresee costs that we have good reasons to avoid and others that we have

[54] *Ibid.*, 24.

[55] This problem is also raised in Neil A. Manson, "Formulating the Precautionary Principle," *Environmental Ethics* 24 (2002): 192–202.

[56] Bjørn Lomborg, *Cool It: The Skeptical Environmentalist's Guide to Global Warming* (New York: Alfred A. Knopf, 2008), p. 160.

less reason to avoid, this is not enough to engage in cost-benefit analysis. The fundamental characteristic of the circumstances in which minimax is justified is that no objective probabilities can be attached to the costs that we have good reason to want to avoid. The charge that no new advice is offered is plausible only if cost-benefits analyses are capable of offering advice. Because, with respect to the cascading uncertainties of climate change, they are not able to provide advice, Lomborg's charge is plausible only if the problem is wished away.

In the course of a useful discussion of catastrophe, Richard Posner contends that the precautionary principle, "[i]n its more tempered versions . . . is indistinguishable from cost-benefit analyses with risk aversion assumed."[57] He proceeds to examine various techniques for "adjusting cost-benefit analysis to reflect the presence of radical, non-quantifiable uncertainty."[58] There are two problems with this approach. First, reliance on "risk aversion" to do the work of supporting a precautionary approach in matters of public policy is completely unsatisfactory. It is an instance of a fallacy that pops up regularly in some economic analyses of public policy. Because I'll have occasion to discuss this again, it is useful to give it a name. Let's call it *the psychological fallacy*. The psychological fallacy is committed when one takes the psychological dispositions of individuals, vis-à-vis their own well-being, as models of the kinds of reasons appropriate for public policy that affects other people.

The concept of risk aversion presumably refers to a person's psychological propensity not to tolerate circumstances in which there exists either a high probability of moderately negative outcome or a low probability of an especially bad outcome. But there is no connection between what a policy maker can psychologically tolerate and what is owed to those affected by the policy. The problem here is with the psychological language. Whether or not a policy maker should take a precautionary approach with respect to climate sensitivity has nothing to do with her risk aversion, but it has a great deal to do with what she owes those who are subjected to very bad, even catastrophic, outcomes if the policy maker bets that climate change is lower than the scientific consensus and is wrong. Perhaps the thought behind the language of risk aversion is that it is the aversions of the population, not the policy maker, that are relevant in this case. But that also cannot be correct because the issue of climate change mitigation has much to do with what the present generation owes future generations, and that cannot plausibly be identified simply as the risk to which the present generation is averse. It is then an utterly implausible

57 Posner, *Catastrophe*, 140.
58 *Ibid.*, 175.

characterization of a "tempered version" of the precautionary principle for public policy purposes to claim that it involves risk aversion.

The second problem with Posner's approach is that it is, on the face of it, simply contradictory. Cost-benefit analyses require estimations of risks and there are no credible estimations of these in cases of "radical, non-quantifiable uncertainty." Therefore, it seems simply absurd to suppose that a tempered version of precaution deals with uncertainty by collapsing into some version of a cost-benefit analysis. Posner at least sometimes wiggles free of the grip of contradiction by – quite sensibly – not employing cost-benefit analyses at all when dealing with uncertainty. Consider the following passage:

> But the possibility of abrupt warming should not be ignored. Suppose there is some unknown but not wholly negligible probability that the $1 trillion loss from global warming will hit in 2014 and that it will be too late then to do anything to avert it. That would be a reason to impose stringent emissions controls earlier even though by doing so we would lose the opportunity to avoid their costs by waiting to see whether they would actually be needed.[59]

This is straightforward minimax reasoning. Two possible courses of action are imagined. In one scenario we pursue mitigation before 2014; in the other, we do not. The worst-case scenario in the first case is that we spend money that was unnecessary. But in the second, the worst that can happen is that abrupt climate change will begin and it will be too late to do anything about it. There is no estimation of probabilities here; no comparison of risks to benefits. Posner reasonably concludes we have a reason for mitigation. And to revert to the thought of the previous paragraph for just a moment, the reason would not be that policy makers or present citizens are psychologically averse to catastrophe. Rather, it must be that we owe it to future generations to incur a small cost in order to avert the distinct, yet indeterminate, possibility of very serious moral loss caused by climate change.

Instead of a precautionary principle, Posner examines four alternative techniques for dealing with uncertainty. These are information markets, inverse cost-benefit analyses, tolerable windows, and risk assessment.[60] The first relies on risk estimates from private forecasters, such as insurers, who have a financial incentive to get it right. The second infers probabilities of bad outcomes by dividing the costs actually incurred to insure against them by the loss that would be incurred if they were to come to pass (using the formula $C = PL$, where C is the cost of prevention, P is the probability, and L is the loss). The

[59] *Ibid.*, 163.
[60] *Ibid.*, 174–187.

third supposes that although we cannot know the amount at which the cost of prevention equals the cost of probability-discounted-loss, we can surmise when the costs of prevention are low in relation to the latter. The fourth compares the risks of the bad outcomes of courses of action without seeking to compare their costs. The first, second, and fourth of these techniques fail in the same way that Lomborg's account fails: They simply wish away the problem that there are no scientifically credible probability assessments in the case of the uncertainty. In the absence of the science, there is no reason to trust the probability assessments of insurers, to believe that government expenditures are based on reliable probability estimates, or to compare the risks of bad outcomes.

Posner's third approach is more promising. He recommends choosing in a manner that avoids outcomes that we have reason to avoid when the costs of doing so seem much less than the cost of the outcome. Assuming we have much less reason to care about the costs of avoidance (even if they turn out to be unnecessary costs) than the costs of the bad outcome, this is simply a credible version of the minimax rule. Posner does not make the assumption explicit, but it is, as we have seen, important for a reasonable application of the rule. It is not, then, that a tempered version of the precautionary principle is indistinguishable from a cost-benefit analysis, as Posner claims. On the contrary, what's left of a cost-benefit analysis in circumstances of uncertainty sometimes makes it reasonable to employ precautionary reasoning based on the minimax rule.

We have seen in this chapter that epistemic uncertainties inherent in climate change forecasts provide good additional reasons for mitigation. If climate sensitivity is 3°C, or higher, and policy makers bet on a lower value, the regulatory regime will subject billions of people to the possibility of food insecurity, possibly allow massive species extinction, and increase the threats of catastrophic change. All of this would have been avoidable at comparatively small costs. Presumably, such costs could be arranged in a way that would not fall heavily on the poor. In this circumstance, uncertainties in existing climate science provide reasons to believe that climate change, left unmitigated, would be very dangerous.

4

Discounting the Future and the Morality in Climate Change Economics

"[A]nd Esau came from the field, and he was faint: And Esau said to Jacob, Feed me, I pray thee, with that same red pottage; for I am faint...And Jacob said, Sell me this day thy birthright. And Esau said, Behold, I am at the point to die: and what profit shall this birthright do to me? And Jacob said, Swear to me this day; and he sware unto him: and he sold his birthright unto Jacob. Then Jacob gave Esau bread and pottage of lentiles; and he did eat and drink, and rose up, and went his way: thus Esau despised his birthright."

– Genesis 25: 29–34

Some readers of *Genesis* believe that in this passage, Jacob exploited the desperately hungry Esau. Others believe that Esau simply acted in a profoundly imprudent manner. The author of the passage seems to be in the latter camp because he judges Essau to have despised his birthright. If that is the case, this ancient story is but a vivid example of a familiar foible. There is considerable evidence that people generally prefer present benefits over greater future benefits.[1] Economists have a word for this; they call it "imprudence." That seems fair enough. Some economists have the view that such a preference should be reflected in the value attached to the future costs and benefits of public policy. But when those costs and benefits are enjoyed by other people, the preference for present goods is not imprudence, but a kind of partiality not obviously different from other kinds of pernicious favoritism, such as using public policy to benefit people to whom one feels especially close.

In Chapter 3, I noted an instance of the psychological fallacy; namely, taking the psychological dispositions of an individual vis-à-vis her own well-being as models of – or reasons for – public policy that affects other people.

---

[1] See, for example, John T. Warner and Saul Pleeter, "The Personal Discount Rate," *American Economic Review* 91 (2001): 33–53.

Climate change policy requires us to think about the future; not only our own future, but also the future that includes the lives and well-being of people who will come to be long after we are dead. Even if we begin reducing CO_2 emissions today, it will likely take decades before atmospheric concentrations of CO_2 stabilize. This is because a great deal of CO_2 stays in the atmosphere a very long time after we emit it. According to the Intergovernmental Panel on Climate Change (IPCC),

> [C]arbon dioxide (CO_2) is exchanged between the atmosphere, the ocean and the land through processes such as atmosphere-ocean gas transfer and chemical (e.g., weathering) and biological (e.g., photosynthesis) processes. While more than half of the CO_2 emitted is currently removed from the atmosphere within a century, some fraction (about 20%) of emitted CO_2 remains in the atmosphere for many millennia.[2]

Our CO_2 emissions accumulate in the atmosphere because the rate of emissions is higher than that by which natural processes remove CO_2, just as water fills a basin if it is gushing out of the faucet more quickly than it is flowing down the sink's drain. Additionally, the process of warming the Earth is a very long one owing to the thermal inertia of the oceans. Even if we were to halt all emissions now, the warming commitment by the end of this century would be an *additional* 0.4–0.6°C (0.72–1.08°F). The forecasted consequence of humanity's cumulative CO_2 emissions to date is an eventual total warming of between 1 and 1.5°C.[3]

The United Nations Framework Convention on Climate Change (UNFCCC) recognizes this intergenerational concern by affirming that, "[t]he Parties should protect the climate system for the benefit of present and future generations of humankind."[4] It would be another instance of the psychological fallacy to deem a failure properly to consider the lives and well-being of people who will come to be in the future to be mere imprudence. As discussed in Chapter 1, unlike in the case of the personal analogy, dangerous climate change is a threat to other-regarding values. The norm violated would not be one of prudence or self-interest, but of morality; of what we owe other persons.

[2] G. A. Meehl, et al., "Global Climate Projections," in S. Solomon, et al., eds., Climate Change 2007: The Physical Science Basis. Contribution of Working Group I to the Fourth Assessment Report of the Intergovernmental Panel on Climate Change (Cambridge University Press: Cambridge and New York, 2007). http://www.ipcc.ch/pdf/assessment-report/ar4/wg1/ar4-wg1-chapter10.pdf (accessed February 20, 2009).

[3] Gerald A. Meehl, et al., "How Much More Global Warming and Sea Level Rise?" *Science* 307 (2005): 1769–1772.

[4] United Nations Framework Convention on Climate Change, 1992. http://unfccc.int/essential_background/convention/background/items/2853.php (accessed October 4, 2012).

In what way, if at all, does it matter that the effects of climate change will be borne by others in the future rather than by the current generation? Does their temporal distance give us a reason to do less? I discuss these questions in this chapter. This involves looking at the practice of discounting future costs and benefits. Economic analyses of projects whose benefits will be enjoyed in the future contain a factor by which future costs and benefits are discounted. This is the social discount rate. Economists employing the social discount rate offer several reasons for such discounting. But the higher the social discount rate applied to the projected future costs of climate change, the stronger the reasons for present consumption and the weaker the reasons for mitigation. Most of the reasons for discounting are not compelling. The concern here is not merely with imprudent choices, such as Esau's. When the lives and well-being of other people are at stake, the concern is that discounting is often morally unjustified. Some of these reasons are part of a package – the widely used conception of intergenerational justice called *discounted utilitarianism*. The arguments in the first several sections of this chapter discuss discounting and discounted utilitarianism. These sections cast considerable doubt on discounted utilitarianism because of the discounting on which it relies. But they also cast doubts on the cost estimates of climate change arrived at by economists who employ this social discount rate, which is the dominant practice in the economics of climate change. The estimates are likely to underestimate future costs and thereby suggest a reason for consumption that does not exist.

In contrast to discounted utilitarianism, when thinking about the costs of future climate change we should rely on the version of precautionary reasoning that I defended in Chapter 3. Unmitigated climate change threatens possible grave outcomes that are uncertain, but generally explicable given current scientific understandings. And given what we do know, circumstances are producing some of the known causal factors that would bring these outcomes about. This adds significantly to the reasons for mitigation and makes the effort to arrive at an international climate change mitigation treaty exceedingly important.

A REASON FOR CONSUMPTION?

The social discount rate is the factor by which, when planning for projects that extend into the future, future costs and benefits are discounted. Suppose that we divide all our income into two groups: that which we plan to spend on consumption and that which we plan to save for future costs. The amount that we should put in the second group depends on an estimate of the future costs. The higher the social discount rate applied to future costs, the less we need to set aside to meet them, and the more income we have available for consumption now. In light of the discussion in Chapter 1, the social discount rate is relevant, then, to the reasons that we have for consumption and mitigation. The less costly the impact of climate change, the greater the reason we have to spend money on problems that confront us now.

One place in the policy literature on climate change where the social discount rate is especially important is in the discussion of the estimated social cost of carbon. The social cost of carbon is the projected cost of the future effects of current CO_2 emissions. It is a measure of the monetary costs of climate change that we pass on for people in the future to absorb. It is no surprise that there is considerable disagreement about the social costs of carbon. In 2006 the British government issued a comprehensive report on the economics of climate change, written under the supervision of Sir Nicholas Stern titled the *Stern Review: The Economics of Climate Change* (the *Stern Review*).[5] The *Stern Review* estimates that the social cost of CO_2 in the absence of climate change mitigation is presently about $85 per metric ton.[6] The cost tends to rise over time as atmospheric concentrations of CO_2 grow and the resulting damages of additional emissions are more than those of previous emissions. In contrast, William Nordhaus, an economist who has been at the forefront of developing complex integrated-assessment models of the costs of climate change, calculates the social cost of CO_2 to be $7.40 per metric ton.[7] In other words, the *Stern Review* takes the effects of our CO_2 emissions to be more than eleven times more costly than does Nordhaus.

The different calculations of the social cost of carbon lead to significantly different mitigation policy recommendations. The *Stern Review* advocates immediate action aiming at significant reductions in CO_2 emissions by 2050 and ultimately reductions amounting to more than 80 percent of current emissions. Nordhaus, in contrast, argues for a more gradual approach. His policy ramp recommends 25 percent reductions by 2050 and only 45 percent by 2100.[8]

Remarkably, the disagreement between Stern and Nordhaus has little to do with differing estimates of climate sensitivity or to disagreements about the kinds of damages that temperature increases will produce. Much more

---

[5] Nicolas Stern, *Stern Review: The Economics of Climate Change*, http://www.hm-treasury.gov .uk/sternreview_index.htm (accessed October 5, 2012).

[6] *Ibid.*, p. xvi.

[7] William Nordhaus, *A Question of Balance: Weighing the Options on Global Warming Policies* (New Haven and London: Yale University Press, 2008), p. 15.

[8] *Ibid.*, p. 14.

contentious is the amount by which costs to future people should be discounted. The higher the social discount rate applied to future costs, the less reason there is to invest in future cost reductions through mitigation or savings for the adaptation needs and the more money there is to spend on present consumption. The social discount rate, in other words, provides an instance of the second kind of reason identified in Chapter 1 – a reason that present people should not forego the benefits of fossil fuels – a reason for consumption.

Economists, of course, have no special expertise in moral questions. Perhaps it is for this reason that in some economic discussions of the social discount rate the question of the appropriate rate is treated solely as an empirical matter. Some say that we should turn to capital markets to discern how much people in fact discount the future. The idea is that when evaluating our reasons for consumption and mitigation, it is appropriate to base decisions on how imprudent people really are, as the discount rate in the financial markets is supposed to reveal. If the market reveals the present value of one's birthright is a bowl of lentils, then who is Esau to disagree? Such an approach reaches the level of high principle in the writings of Nordhaus. He contends that the social discount rate applied to the future costs of climate change should approximate "the *real* real interest rate – as the benchmarks for climate investments. The normatively acceptable real interest rates prescribed by philosophers, or the British government are irrelevant to determining the appropriate discount rate to use in the actual financial and capital markets of the United States, China, Brazil, and the rest of the world."[9] Moreover, in criticizing the *Stern Review's* account of the social discount rate, Nordhaus asserts that "it takes the lofty vantage point of the world social planner, perhaps stoking the dying embers of the British Empire, in determining the way in which the world should combat the dangers of global warming."[10] This anti-(British) imperialism and suspicion against philosopher kings is a lot of decoration on behalf of a homely reason for consumption.

This is a good time to return to the lesson of the personal analogy in Chapter 1. There the idea was that the normative question of what risks one *should* assume is underdetermined by both the empirical forecasts of natural scientists and the behavioral summaries of the social scientists. The risks that one should assume depend crucially on what one values. That was for matters of prudence. Decisions about climate change-related risks and uncertainties are not matters of prudence. Appreciation of the psychological fallacy steers us

---

9   William D. Nordhaus, "A Review of the Stern Review on the Economics of Climate Change," Journal of Economic Literature XLV (2007): 692.
10   Nordhaus, *Balance*, p. 174.

clear of thinking that what we owe other persons is decided by our preferences – how much we, in fact, discount future costs. What we prefer for ourselves, as expressed in our investment decisions, tells us nothing about what our moral duties are to persons who will follow us.

The question of the rate by which the future costs of climate change should be discounted has an inescapably moral character. This is not to argue that there are no economic considerations that go into setting the rate. As with many matters of public policy, good policy requires analysis from many areas of expertise. Economists are well equipped to forecast costs and to aggregate them. But insofar as the social discount rate is relevant to whether we have a reason to consume or to mitigate, assessing its proper value necessarily includes moral argumentation. Here the psychological fallacy leads us astray.

DISCOUNTED UTILITARIANISM

In a great many economic analyses of how much we should invest in climate change mitigation or save for adaptation, a particular understanding of the aims of investment is employed. This is the view of discounted utilitarianism. The utilitarian moral view evaluates actions or policies according to how much they contribute to the overall welfare of those affected by them. The affected parties are typically understood to include not only those who are actually affected by the action pursued, but also those who might have been affected if instead another action had been pursued (without a distinction between commission and omission) because the point is to recommend an action from a menu of available actions.[11] Utilitarianism recommends that action or policy maximally contributes once all the contributions and negative effects are aggregated. Utilitarianism incorporates, at a fundamental level, a particular understanding of impartiality. This is typically expressed as the principle – attributed to Jeremy Bentham – that each person is to count for one and no one for more than one.[12] In the classical theory of utilitarianism,

---

[11]   That we have responsibility to those whom we do not affect (but might have affected if we acted differently), but who are basically identical in kind to those whom we do affect, is part of the doctrine of called "negative responsibility" by Bernard Williams. Williams offers a trenchant criticism of the doctrine in his "A Critique of Utilitarianism," in J. J. C. Smart and Bernard Williams, *Utilitarianism: For and Against* (Cambridge: Cambridge University Press, 1973), p. 95 ff.

[12]   John Stuart Mill refers to this as "Bentham's dictum" without citation in *On Liberty*, book 5. An editor of Mills's work, J. M. Robinson (in *Essays on Religion, Ethics, and Society* [Toronto: University of Toronto Press, 1969], p. 257) cites Bentham's *Plan of Parliamentary Reform* in *Works* vol. III, p. 459. http://files.libertyfund.org/files/1922/0872.03_Bk.pdf (accessed October 29, 2012). The reference might be to the following: "The happiness and unhappiness of any

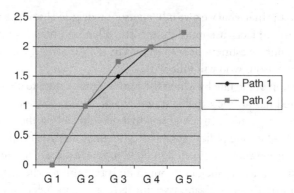

FIGURE 4.1. Two hypothetical intergenerational consumption paths differing at one generation.

a unit of happiness or pleasure counts for one and only one, regardless of the person (or, more inclusively, regardless of the sentient being); these are aggregated for, and across, persons. Modern economics shies away from trying to compare and aggregate subjective experiences, and substitutes optimizing the satisfaction of preferences – as revealed by market choices – as the goal of policy. Intergenerational utilitarianism takes this process of objectification one step further. Because future preferences cannot be revealed, it seeks the policy that will optimize consumption over an infinite time horizon. Over time, happiness has transmogrified into consumption.

Consider Figure 4.1 where the horizontal axis represents generations and the vertical global GDP. (No attempt at a real representation of economic growth over time is intended.) If the global GDP for a generation is used as the measure of consumption, then intergenerational utilitarianism recommends the policy-producing Path 2 over the policy-producing Path 1 because Path 2 has at least one generation with higher consumption (and none with lower) than Path 1.

The comparison in Figure 4.1 is, of course, idealized. In fact, projections need to compare multiple consumption paths (produced by multiple policies), which vary at several points. Figure 4.2 simplifies this by representing two different consumption paths, which vary at two points. Utilitarianism recommends the policy that has the highest aggregate consumption over an infinite time horizon. This would be represented by the curve with the greatest area below it.

one member of the community – high or low, rich or poor – what greater or less part is it of the universal happiness than that of any other?" In any case, the uncited attribution of this principle to Bentham is by now a commonplace.

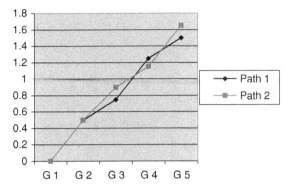

FIGURE 4.2. Two hypothetical intergenerational consumption paths differing at two generations.

In Chapter 1 I argued that intergenerational utilitarianism is inconsistent with equal respect for human dignity because, in principle, it would require imposing significantly inferior life prospects on a person (or an entire generation) on grounds that it leads to minor improvements to people over an infinite time horizon. In other words, intergenerational utilitarianism would require policy choices in which persons in a particular generation are used merely as a means to support improvements for people in other generations. My present focus is not on that significant moral defect, but on the manner in which intergenerational utilitarianism requires discounting some of the future costs of climate change.

There are three factors in the standard social discount rate employed by economists, and each serves a different purpose in assessing optimal consumption. I set details of these factors aside until the next section in order to consider an intuitive reason to employ a social discount rate if the project is to optimize consumption over an infinite time horizon. In the absence of discounting future benefits, a sacrifice of consumption of any size in the present will be justified if it leads to a benefit – again of any size – into the infinite future because even an enormous sacrifice in the present will be outweighed by a miniscule benefit summed infinitely into the future.[13] One reason to employ a social discount rate, given the goal of optimizing consumption

[13] This argument for discounting originates in Tjalling C. Koopmans, "Stationary Ordinal Utility and Impatience," *Ecomentrica* 28 (196): 287–309, esp. 306. It is made in an intuitive way by Kenneth J. Arrow, "Discounting, Morality, and Gaming," in Paul R. Portney and John P. Weyant, eds., *Discounting and Intergenerational Equity* (Washington, DC: Resources for the Future, 1999), p. 14. It is critically discussed by John Broome in *Counting the Costs of Climate Change* (London: The White Horse Press, 1992), pp. 104–106.

intergenerationally, is that absent discounting future consumption gains, the massive losses of an earlier generation would be offset by the tiny gains of a sufficient number of future generations.[14] I call this *the intuitive defense*.

The intuitive defense has wide currency in economic discussions of discounting. To cite just one example, Nordhaus employs the intuitive defense in his climate-wrinkle experiment:

> Suppose that scientists discover a wrinkle in the climate system that will cause damages equal to 0.1 percent of net consumption starting in 2200 and continuing at that rate forever after. How large a one-time investment would be justified today to remove the wrinkle that starts only after two centuries? If we use the methodology of the *Stern Review*, the answer is that we should pay up to 56 percent of one year's world consumption today to remove the wrinkle. In other words, it is worth a one-time consumption hit of approximately $30,000 billion today to fix a tiny problem that begins in 2200.[15]

In this quotation Nordhaus is criticizing the generally low social discount rate of the *Stern Review*, not any particular factor of the rate. Optimizing across infinite future generations with a negligible social discount rate threatens the present generation with overwhelming burdens.

The problem that the intuitive defense responds to is not peculiar to optimizing over infinite future time horizons. Rather, the problem is an instance of a general feature of utilitarian reasoning.[16] Even within the time frame of a single generation, happiness maximization would seem to require wealthy individuals to give extraordinary amounts – up to the point of equalizing marginal utility – if there are a great many people suffering from extreme deprivation. Consider Peter Singer's non-intergenerational account of our duties to relieve suffering:

---

[14] See also Henry Shue, "Bequeathing Hazards: Security Rights and Property Rights of Future Humans," in Mohamed Dore and Timothy Mount, eds., *Limits to Markets: Equity and the Global Environment* (Malden: Blackwell Publishers, 1998), p. 49.

[15] Nordhaus, *Balance*, p. 182. Derek Parfit construes the social discount rate as expressing the belief "that the importance of future benefits steadily declines" in *Reasons and Persons* (Oxford: Clarendon University 1984), p. 484 and therefore rejects the intuitive defense. But this construal is too narrow. It is accurate with respect to the element of the social discount rate known as the rate of pure time preference. If Parfit's point is that the intuitive defense does not support the rate of pure time preference specifically, that seems correct.

[16] This is referred to as the tyranny of aggregation in Marc Fleurbaey and Bertil Tungodden, "The Tyranny of Non-Aggregation versus the Tyranny of Aggregation in Social Choices: A Real Dilemma," *Economic Theory* 44 (2010): 399–414. According to the intuitive defense, discounting mitigates the tyranny of imposing huge costs of present generations for the sake of small benefits aggregated over an infinite time horizon.

[I]f it is in our power to prevent something bad from happening, without thereby sacrificing anything of comparable moral importance, we ought, morally, to do it. By "without sacrificing anything of comparable moral importance" I mean without causing anything else comparably bad to happen, or doing something that is wrong in itself, or failing to promote some moral good, comparable in significance to the bad thing that we can prevent.[17]

The first sentence of the quotation expresses a principle that might be characterized as negatively utilitarian. We are to minimize the aggregate bad, rather than to optimize overall goodness. In pressing this principle, Singer is well aware that its observance would alter "the way of life that has come to be taken for granted in our society."[18] The difference between Singer's principle and the demands on the present generation made by optimization over an infinite time horizon is that the extraordinary giving required by Singer's principle is contingent on the extent and depth of current deprivation whereas intergenerational optimization creates overwhelming savings demands that are not all contingent. Rather, they are a necessary feature of optimizing over an infinite time horizon. The intuitive defense is an attempt to lighten the burden of this necessary feature optimization over an infinite time horizon.

One might read the wrinkle experiment as a criticism of intergenerational utilitarianism of the kind that I expressed in Chapter 1; namely, that the basic moral defect of optimizing across generations is that it treats persons and generations – in the wrinkle experiment, members of the present generation – merely as a means to the goal of producing the greatest possible amount of impersonal good – in this case, consumption. I trust that the critique offered in Chapter 1 rings intuitively clear to many people who think carefully about the moral merits of intergenerational utilitarianism. Instead of this, Nordhaus and many other economists seek to introduce discounting future benefits as a means to avoid laying heavy burdens on the present for the sake of the future. As we shall see presently, this approach, so common in economics, strays from the cause of impartiality defended by Bentham and is incapable of understanding moral judgments regarding inequality.

THE SOCIAL DISCOUNT RATE

The conception of the social discount rate that I am interested in here goes back to the pioneering work of the mathematician cum philosopher cum

---

[17] Peter Singer, "Famine, Affluence, and Morality," *Philosophy and Public Affairs* 1 (1971): 231.
[18] *Ibid.*, 230.

economist, Frank P. Ramsey.[19] Ramsey discusses the social discount rate within a classically utilitarian framework. His goal is to determine the proper rate of savings for a single agent in order to achieve optimal utility over time, or what he calls "Bliss." Savings are disutilities experienced for the sake of future increases in utility.[20]

There are several important differences between Ramsey's discussion and contemporary discussions. As I have already noted, in the current literature, consumption rather than subjective utility is the object of maximization.[21] Also, in contemporary discussions, the consumption of multiple persons in a generation is averaged. And to avoid problems associated with population growth and changing preferences for consumption, identical individuals are usually assumed. Perhaps most important of all, but unnoticed by a great many economists, is that in Ramsey's account the social discount rate is a normative value. It is derived from three factors: the elasticity of marginal consumption, the rate of pure time preference, and the rate of growth – each of which is independently justified.[22] In contrast, in a doomed effort to avoid normative analysis, many contemporary economic discussions of the social discount rate for climate change take the social discount rate as given empirically by the rate of return on long-term savings and seek instead to derive the factors that comprise it. This effort errs because the fundamental question that the social discount rate is meant to address is normative: What is the rate by which future costs *should* be discounted? Analyses drawing on Ramsey's approach recognize

---

[19] The classical account of the social discount rate employs what has come to be called the Ramsey equation. For the source of the equation, see F. P. Ramsey, "A Mathematical Theory of Saving," *The Economic Journal* 38 (1928): 543–559. The equation is much discussed in the economics literature. For a good discussion, see Partha Dasgupta, "Discounting Climate Change," *Journal of Risk and Uncertainty* 37 (2008): 141–169. There are many modifications and criticisms of the equation in the economics literature. An alternative overlapping generations model (not employing the Ramsey equation) is developed in Richard B. Howarth, "An Overlapping Generations Model of Climate-Economy Interactions," *The Scandinavian Journal of Economics* 100 (1998): 575–591.

[20] Ramsey, "Savings," 545.

[21] This move from utility to consumption – but not the moral problems that I discuss in section IV – is discussed in G. M. Heal, "Intertemporal Welfare Economics and the Environment," in Karl-Göran Mäler and Jeffrey Vincent, eds., *Handbook of Environmental Economics*, vol. 3 (Amsterdam: Elsevier, 2005).

[22] A canonical statement of the Ramsey equation is $\rho_t \approx \delta + \eta g(C_t)$. The left-hand symbol represents the social discount rate applied to future benefits and losses at a certain time, usually the present. δ stands for the rate of pure time preference, η for the elasticity of marginal consumption, and $g(C_t)$ for the rate of economic growth from the particular time (growth as measured by growth in consumption). Ramsey takes ρ_t to depend on the values of δ, η, g, which are independently ascertained.

three factors in discounting the present value of future costs. I discuss these in each of the following three sections.

THE ELASTICITY OF MARGINAL CONSUMPTION

Looking back over the past eighty years, the global economy has grown considerably. Growth produces wealth. Inflation adjusted per capita GDP in the United States in 1931 was less than $7,000. In 2011 it was more than $48,000. If we are fortunate to live in countries that have experienced this growth, we are wealthier now than our grandparents or great-grandparents were in 1931. We are better educated and less threatened by poverty, disease, and unsafe work conditions. If we set aside climate change-induced uncertainty about economic growth and its regional variation, and instead assume average economic growth continues over the medium term, we have reason to believe that on average, people in eighty years will be considerably wealthier than we are now. This, obviously, could turn out to be false if there were climate change-related disasters.

When considering spending money to improve the condition of people eighty years in the future, how much should our choice be sensitive to their assumed better circumstances? More precisely, how should we value equal marginal increases in wealth for richer (future) people in comparison to poorer (present) people? In plain language, should our money go to us or to them? To the classical utilitarian project of maximizing welfare, this is a very important question. For if a given increase in wealth brings more happiness to a poor person than to a richer one, the project of maximizing happiness would go awry if that increase were assigned to the rich rather than the poor person. For the classical utilitarian, the question of how to distribute wealth is always decided by the goal of happiness maximization.

If one knew enough about a particular person (and if there were a noncontroversial means of measuring happiness), one might be able to determine exactly how happy a deposit into the person's bank account might make that individual. But that would be unrealistic when speaking at the level of generality that policy requires. And we would, in any case, confront the problem of how to compare increases in happiness between persons. So, utilitarians rely on a rough generalization that one can attempt to refine as the circumstances allow. The generalization is called *decreasing marginal utility*. Although rough and a generalization, the idea is assumed to have a basis in psychological fact. As a general rule, the more a person has of a particular good, the less happiness an additional unit of the good brings. To see this, think only of how much you are willing to pay for the first cup of coffee in the morning, as opposed to the second, or the third . . . or the sixth.

Because utilitarianism takes our sole fundamental moral duty to be happiness maximization, it is exceedingly important to take note of the general fact of decreasing marginally utility. Failing to do so could result in distributions that are not happiness maximizing, and therefore morally inferior. It is important to appreciate that for the classical utilitarian; determining whether the deposit should be made into the account of the poor or rich person is not based on the preferences of the depositor to achieve a more equal distribution, or on his aversion to inequality. There is a sense in which it is based on the preferences of the receiver of the deposit because the idea is that the poor person will be made happier by the deposit than the rich person. But this preference – if that's what it is – is not for equality. The assumption is not that the poor person is made happier knowing that the distribution is more egalitarian, but that the individual is made happier having the deposit, which incidentally involves the distribution being more egalitarian. How egalitarian a utilitarian should be depends on the features of the general fact of declining marginal utility and how much inequality can establish incentives that promote economic activity, which improves overall happiness.

The elasticity of marginal consumption is the conceptual descendent of decreasing marginal utility after happiness has mutated into consumption. The basic idea is that consumption to the rich is less valuable than consumption to the poor because the rich consumer has more.[23] So, on the assumption of economic growth, taking climate change costs as decreases in consumption to richer future generations and mitigation investments as decreases in consumption to the relatively poorer present generation, a unit of decreased consumption for the present generation would be worse than the same unit decrease for the future one. If this is the case, then the present value of future climate change losses is less than the present value of the same amount of present losses. Therefore, our mitigation investments should not cover the value that a particular decrease in consumption would have to us, but the lesser value it would have to future wealthier people. To do otherwise would be to fail to optimize consumption. In other words, for a decrease in consumption of X units (either to us or to people in the future), which has a monetary value $M, we should presently only pay $M – d. Here, "d" represents the

---

[23] If $g(C_t) > o$ and $\eta > 1$, then the rate of growth is positively multiplied in the social discount rate to account for the inverse relationship between the value of a marginal increase in consumption and the marginal growth of consumption. Assigning a value to $\eta > 1$ adds a factor to the discount rate that is meant to consider the lower value of growth to future wealthier persons, not merely its effect on the prices of commodities.

difference in value of the consumption for the two generations because of the later generation's relative affluence.

The value assigned to the elasticity of marginal consumption is supposed to be the ratio by which the value of the extra unit of consumption decreases because, and as, a generation's overall consumption increases. The *Stern Review* assumes that the value does not decrease at all just because the future generation is richer.[24] In contrast, Nordhaus assumes that the value of consumption to the rich is significantly less than to the poor because the rich are richer. Nordhaus holds that the rate of discounting should be double the rate of growth in order to account for the differential value of growth to the rich and the poor.[25] Nordhaus is then significantly more egalitarian than the *Stern Review* in this regard.[26]

The economist Partha Dasgupta illustrates the significance of Nordhaus's approach as follows:

> Consider two people, 1 and 2 . . . whose annual consumptions (at purchasing power parity) are $360 and $36,000 respectively. The former is below even the World Bank's "dollar-a-day" person, while the latter is well above the annual income of the average resident of the European Union, which is approximately $29,000 . . . [The approach advocated by Nordhaus holds] a 50% decrease in person 2's consumption to be ethically equivalent to a 1% decrease in person 1's consumption.

For Dasgupta, the point of this illustration is to guide our moral intuitions. This is important to Dasgupta because, unlike the classical utilitarian, he thinks that ultimately the degree to which consumption should be discounted on the basis of wealth is a matter settled by moral intuitions. Despite the huge hit to person 2, Dasgupta rightly cautions against assuming that a view such as Nordhaus's is counterintuitive without taking heed of the dynamic context.

> If this trade-off feels unreasonable, we should ask whether the thought experiment we are conducting is itself reasonable. After all, our attitude toward income transfers is influenced not only by our concern for equality of outcome, but also by the recognition that incentives matter. Incentives in turn are shaped by the presence of moral hazard and adverse selection. Our

---

[24] *The Stern Review* assigns η the value of 1. See *The Stern Review*, 46. Since then, Stern has expressed a more egalitarian leaning, taking η as either 1.5 or 2. See Nicolas Stern "The Economics of Climate Change," *American Economic Review* 98 (2008): 1–37.

[25] Nordhaus assigns η the value of 2. See Nordhaus, *Balance*, p. 61.

[26] More egalitarian still is Partha Dasgupta, who speculates that the range for an appropriate η is 2–4. See Dasgupta, "Discounting."

thought experiment is oblivious of incentives, it is of little value in testing our intuition.[27]

If the point is to maximize growth over time, then incentives for growth would surely matter. We could not, then, simply look to the immediate outcome of assigning costs to the poor or the rich. We would have to consider the dynamic effects of such assignments.

More important for our purposes, however, is the manner by which Dasgupta's approach differs from the classical utilitarian concern of decreasing marginal utility. His intuitive approach is meant to test what "feels unreasonable," rather than to test a factual claim about what would satisfy the goal of optimizing utility given a psychological fact about people. For classical utilitarians, factoring in decreasing marginal utility serves the goal of optimizing happiness. For Dasgupta, the correct value of the elasticity of marginal consumption is entirely independent of the goal of optimizing.

Unlike identifying decreasing marginal utility in the classical utilitarian project of happiness maximization, discussions of the elasticity of marginal consumption in climate change economics cannot assume that there is a fact of the matter that the elasticity should track. Of course, for the classical utilitarian this is often a crude psychological generalization about the relationship between how much happier a person is made by an additional unit of a good and how much the person already possesses. But there is at least the presumption that the answer does not depend on the intuitions or the preferences of the policy maker. For Dasgupta, in the grips of the psychological fallacy, the elasticity of consumption is based on "inequality aversion." This leads him to find an appropriate value for the elasticity of marginal consumption not by estimating differential utility boosts to the rich and the poor, but by gauging our reactions to the effects that various values of the elasticity of marginal consumption would have on the rate of savings and by considering actual consumer savings behavior.[28] The point of these exercises is, simply, to better understand our aversions to inequality.

The psychological fallacy has gone viral among economists discussing this matter. When Martin Weitzman speaks of the ethical value of the elasticity of marginal consumption, he fastidiously places quotation marks around the word "ethical," perhaps to note a certain ironic usage. He also refers to the elasticity of marginal consumption as a "taste-parameter." Now, if the problem of inequality is merely that it is distasteful, then how much is permissible is

---

[27] *Ibid.*
[28] *Ibid.*

basically the same kind of question as determining whether zinfandel is better than pinot noir or vice-versa. If the matter is merely one of taste, there is no real argument to be had. The only criticism one could make of an unequal distribution – if one can call it a criticism – is that it is out of line with what people generally prefer. Weitzman, like Dasgupta, takes the appropriate value of the elasticity of marginal consumption to be related to observed behavior.[29] We arrive at it by observing how people's preferences for inequality are revealed by their actions, just as the market determines whether more zinfandel than pinot noir should be produced.

Given the utilitarian origins of the theory of optimal intergenerational consumption, the tendency to take judgments of the appropriate value of the elasticity of marginal consumption as largely subjective is surprising. For the classical utilitarian, the rate at which utility declines is based on putative psychological facts about the marginal utility of an action or outcome. Getting these facts right is essential to the goal of overall utility optimization. The required action is a theorem derived from the norm of optimization and the supposed psychological facts of decreasing marginal utility. The person who fails to appreciate the psychological facts of the people affected will not optimize utility.

An explanation for the divergence between these contemporary economists and classical utilitarianism on this point is illustrative of the problematic nature of the optimizing paradigm that many contemporary economists endorse. The divergence derives from the project of maximizing consumption rather than utility. For classical utilitarians, utility is a psychological state assumed to be caused by actions and possessions. According to the psychology implicit in the classical account of decreasing marginal utility, this psychological state has an elastic relation to the actions and possessions that cause utility. For every additional pleasurable action or good thing possessed, the state improves a little less. When the object of value becomes consumption itself rather than the utility produced by consumption, the relationship between a psychological state and an action necessarily disappears. And to state the obvious, without a relation between the psychological state and the action or possession, there can be no psychological fact about the elasticity of the relation. The elasticity of marginal consumption is the mere shell of decreasing marginal utility that remains after the inner psychological state has been excised from the concerns of the theory. As a result, economic discussions of the elasticity of marginal consumption ring hollow. There is no objective relationship that can give them substance. And the necessarily subjective account of moral judgments

---

[29] Weitzman, "A Review," 709.

about inequality that results does nothing to recommend the moral project of seeking the optimal path of average generational consumption. Judging what we owe to other people is not like preferring one wine over another. If some economists think otherwise, this gives us a good reason to be especially suspicious of their advice.

This section has argued that the problems with one factor of the social discount rate run deep. That there can be no objective basis for assigning a value to the elasticity of marginal consumption is a serious strike against the project of optimizing intergenerational consumption insofar as the latter employs a social discount rate that has a factor for the elasticity of marginal consumption. The next section continues this tact with a discussion of the rate of pure time preference.

THE RATE OF PURE TIME PREFERENCE

If you ask whether a person prefers to receive $100 now or $110 next year, chances are that, like Esau, there will be a preference for the lesser amount now. One might have very good reasons for wanting the $100 now rather than next year: One's reputation might raise doubts about a payment, a medical diagnosis might raise doubts that the person will be alive next year to receive the payment, inflation might sufficiently reduce the value of the sum in the future, or one might reasonably expect a greater income in the future. None of these reasons would derive merely from the fact that the payment would occur now instead of next year. In each case there is something additional that makes getting the money now more attractive than getting it later. But suppose that all of these matters (and all matters other than the time difference) could be controlled by, say, a certificate of deposit in a bank now, a physical exam that issues a good bill of health, and mechanisms to index the payment to inflation and to make it proportional to other income, and so on. If, with all of these assurances, one still preferred the lesser cash now, then the preference would be an example of *a pure time preference*; a preference for the cash now for no other reason than that one would have it now rather than later.

The rate of pure time preference in economics is a social, rather than individual, preference. It is the rate at which we prefer consumption now over future consumption simply because it is present consumption, regardless of whether there will be greater wealth in the future. Now, consumption is something that people do, and over a long enough period of time there will be different people doing the consuming. So the rate of pure time preference often expresses a preference for our consumption over investment for the sake of consumption by other people in the future. To put it that way, of course, expresses in glaring fashion the psychological fallacy: How much people *prefer* to save for the benefit of future people is irrelevant to how much they *should* save. So, in the discussion that follows, I consider whether there is any *reason* to believe that what we owe people in the future is less – we should invest less in mitigation projects – simply because they are in the future, regardless of what we might prefer to do.

The rate of pure time preference is often cited as the major source of disagreement between Stern and Nordhaus regarding the social discount rate,[30] even though, as I pointed out in the previous section, they also disagree about the value to assign to the elasticity of marginal consumption.

The science of climate change draws our concern to the effects of our policies far into the future. Can the sheer fact that these effects are on people in the future make them matter less morally? Our moral thinking has developed over a few millennia in circumstances in which our primary moral concerns were directed originally to kith and kin and later to strangers deemed sufficiently similar, or relevantly connected. In the second half of the twentieth century, moral philosophers began to think carefully about moral duties over great geographical distances. A seminal piece in this effort is Peter Singer's article from which the previous quotation was drawn. Singer argues that, all other things being equal, geographical distance is morally irrelevant, at least regarding our duty to help persons in need:

> It makes no moral difference whether the person I can help is a neighbor's child ten yards from me or a Bengali whose name I shall never know, ten thousand miles away . . . I do not think I need to say much in defense of the refusal to take proximity and distance into account. The fact that a person is physically near to us, so that we have personal contact with him, may make it more likely that we shall assist him, but this does not show that we ought to help him rather than another who happens to be further away. If we accept any principle of impartiality, universalizability, equality, or whatever, we cannot discriminate against someone merely because he is far away from us (or we are far away from him).[31]

Two things are worth noting about this passage. First, Singer's condemnation of discrimination in the final sentence is not a rejection of preferring the neighbor's child, if everything else is equal. The sentence that precedes it asserts only that the proximity of the child does not provide an additional moral reason for helping – a reason in addition to neediness that would tilt the moral case in her favor. A preference for the person nearby is permissible

---

[30] See, for example, Dasgupta, "Discounting."
[31] Peter Singer, *Philosophy and Public Affairs* 1 (1971): 231–232.

if there is otherwise a moral equilibrium between the two options. However, a moral requirement of preferring those to whom we are geographically close would be inconsistent, Singer is claiming, with moral impartiality; it would require arbitrary discrimination. Second, Singer is not here making a claim applicable only internally to a utilitarian calculus. Rather, he is making a general claim, regardless of the normative ethical justification of the duty to assist. Singer is asserting a general principle of impartiality in assistance across distance.

Singer's view is that the moral irrelevance of distance is simply an extension of other more familiar interpretations of equality, such as a person's sex, race, or religion that cannot be grounds for discrimination. He then proceeds to argue that the extent of suffering in poor countries allows us to be more efficient in the use of resources that we devote to helping, and this is a morally relevant reason to give more to help poverty that is distant from us. Recalling our discussion of decreasing marginal utility, we are able to help the deeply impoverished more with the same amount of giving.

Now consider the final sentence of the previous quotation, but this time read it as if he means temporal – rather than geographic – distance: "If we accept any principle of impartiality, universalizability, equality, or whatever, we cannot discriminate against someone merely because he is far away from us [in time]." This revised sentence states a principle of impartiality across time. A difference between this principle and the one that Singer employs is that it is difficult to consider examples as lively as facing the choice between helping the neighbor child and the starving Bengali child. In part, this is because in the case of future distress, our efforts are directed to investments that further prevention rather than assistance. Thus, we must consider examples of helping someone now in distress and preventing someone from becoming distressed in the future. The principle stating impartiality across time would hold that, everything else being equal, there is no moral reason to assist in the present. A rule governing how we use our income that requires us to discount the benefits that we can bring to future persons, simply because they are in the future, runs afoul of such impartiality. It also contradicts Bentham's principle of equality: namely, that each person counts for one and no more than one.

Perhaps, however, we have reason for devoting less to hardships of future persons because future hardships are merely probable, not certain. In contrast, we can and do know much about present hardships. Global warming will have climatic consequences, but we cannot be certain how extensive they will be even though we can be confident that, generally, they will include significant hardships caused by drought, flooding, storms, and disease. Of two possible equally bad outcomes, it might make good moral sense to seek to prevent the more likely from occurring. But, this does not gainsay the principal of impartiality across time because it is probability (not temporal location) that provides the reason for discounting. It serves merely to discount the negative value of the outcome by the probability of the outcome – to transform the cost to a risk. It is not to discount the bad outcome simply because it is in the future.

Another criticism of the principle of impartiality across time is that it is simply question begging. Of course, impartiality is inconsistent with discrimination of persons on the basis of temporal distance, but the underlying issue is whether one should be impartial with respect to present and future persons. In response, it is important to observe that impartiality is most important to us in the arena of public policy, which is the arena in which the present value of future climate change costs are established. How much we invest by means of climate change mitigation will largely be a matter of public policy. We take it as an established principle of public policy that invidious discrimination against persons on the basis of their sex, race, or religion is unjust. The prohibition against invidious discrimination is not merely a matter of preference or taste, but a fundamental principle of democratic public life. Administrators, or even popular majorities, cannot disregard it without committing an injustice. The prohibition of discriminating against persons on the basis of their sex, race, or religion extends naturally to proscribing discrimination against them on the basis of their temporal location as well.

Stern is in agreement with this reasoning. He believes that the rate of pure time preference should tend toward zero. Moreover, the *Stern Review* is clear that this is a moral judgment. In a subsequent discussion of the position taken in the *Stern Review*, Stern puts the matter in the following way: "If you stand with your grandchild who let us suppose is 60 years younger than you, can you look her or him in the eye and say 'Well you're coming 60 years after me, so you only get half of what I get?'"[32] The *Stern Review* employs a rate of 0.1, not because Stern believes that the consumption of people in the future is less valuable than that of people in the present, but because he believes that there is a very small chance of mass extinction of the human species by meteor or nuclear war, and because he is calculating into the indefinite future, he wants to account for the possibility of extinction.[33] Technically, this is a risk factor rather than a pure time preference.

In contrast, Nordhaus – in his recent work – employs a rate of pure time preference of 1.5.[34] Nordhaus's attempts to justify temporal discrimination

---

[32] Nicholas Stern, "The Economics of Climate Change," *FST Journal* 19 (2007): 13.

[33] The Stern Review, 46–47.

[34] Nordhaus, *Balance*, p. 61.

are not at all convincing. One argument that he offers is that the overall social discount rate should approximate the real return on capital, which he estimates to be about 5.5 percent. Employing the Ramsey equation (which takes the social discount rate as the sum of the rate of pure time preference added to the product of the rate of growth multiplied by the elasticity of marginal consumption), if growth is projected at 2 percent and the elasticity of marginal consumption is two, the pure time preference must be 1.5 percent to get a social discount rate of 5.5 percent.

There are, I think, two charitable ways to interpret Nordhaus's point about the importance of the rate of return on capital. One is as a claim about the putatively democratic – or perhaps, better, antielitist – nature of using the real return on capital as a normative regulator of the social discount rate. The other is as a claim about the dangers of getting the opportunity costs of investment in climate change mitigation wrong if the real return on capital is not approximated by the social discount rate. It is noteworthy, however, that neither of these arguments actually supports Nordhaus's point about the rate of pure time preference. Because the other variables of the social discount rate could be manipulated to match the social discount rate with the real return on capital, neither of the arguments requires that the rate of pure time preference be anything other than zero. Worse yet, not only do these arguments do nothing to support Nordhaus's claim about the value for the rate of pure rate preference, they also do not even make compelling cases for taking the real rate of return as normative for the social discount rate.

Consider first the "antielitist" argument. The real rate of return on capital is the result of a great many negotiations by individual market agents – lenders and borrowers – all seeking to advance projects for profit. Who knows better than the agents involved about the various considerations relevant to assessing whether an investment will be profitable? When public policy disregards these rates by discounting the future costs at rates that are lower than market rates, more money will go toward future investments than market agents deem correct. Perhaps this is a kind of egregious elitism on the part of philosopher-king policy makers.

Two responses to this argument are in order. First, the normative lesson to be drawn from market investment rates in general is not necessarily that they *ought* to guide public policy.[35] And second, in the case of climate change in particular, we have good reason to worry that the market investment rates undervalue the well-being of future persons. With respect to the first and more

[35] See also Parfit's response to the democratic argument – where it is democratic decisions, not market rates that are supposed to be normative – in *Reasons and Persons*, pp. 480–481.

general point, consider the following case. Suppose that a thorough investigation of home lending practices revealed that mortgage rates systematically discriminated against people on the basis of race, after controlling for all of the relevant credit risk factors.[36] Should we conclude that the normative lesson is to follow the wisdom of the market or is it rather that the market must be regulated to prevent invidious discrimination? To ask the question is to see what a cruel joke the first answer would be. The second point can be appreciated by simply reminding ourselves that climate change is a massive externalities problem, the general character of which is that costs are passed down to future generations. In other words, climate change itself provides some evidence of a general tendency to value the well-being of future generations inappropriately in decentralized market transactions.[37] Such transactions, then, are a poor guide to the appropriate investment rate for mitigating future costs.

What about the second interpretation of Nordhaus's view – namely, that the opportunity costs of investing in climate change mitigation will not be properly appreciated if the discount rate is lower than the real rate of return? Certainly a larger future climate change bill would make the opportunity costs of investing to mitigate higher than a smaller bill because investing to mitigate would take more money away from present consumption. But these higher opportunity costs are only inappropriate if the smaller bill – based on a higher social discount rate – sets the norm for opportunity costs. And that assumes that markets properly capture the future costs of climate change. We would not, however, be discussing adding costs to CO_2 emissions if we believed that.[38] Employing market-established opportunity costs when markets fail to internalize the costs of CO_2 emissions as normative is unjustified. Such externalities also diminish the plausibility of the claim that higher opportunity costs necessarily lead to intergenerational inefficiencies.

Nordhaus just misses the moral nature of the question of what we owe future generations. He reverses the normative order of the derivation of the social discount rate that goes back to Ramsey. He uses an empirical rate – the real rate of return on capital – as determinative and thereby fixes the range for the other normative variables: the elasticity of marginal consumption and the pure rate

[36] This example is, unfortunately, not entirely fictional. Recently the New York State Attorney General's Office and the New York State Department of Banking conducted an investigation of discriminatory lending practices. The investigations lead to settlements with at least two banks. See Bob Tedeschi, "Safeguarding Against Loan Discrimination," *The New York Times*, January 23, 2009. http://www.nytimes.com/2009/01/25/realestate/25mort.html?scp=5&sq=home%20lending%20racial%20discrimination&st=cse (accessed May 7, 2009).

[37] See also Broome in Counting the Costs of Climate Change, pp. 90–92.

[38] See similar argument made by Dasgupta, "Discounting," sec. 3.4.

of time preference. In contrast, the moral question of how much to discount in order optimize positive outcomes over time is more plausibly derived from the assessment of the variables that comprise the social discount rate. The social discount rate is not an empirical fact; it is a tool in the normative project of optimizing.

Weitzman, with somewhat less rhetorical flourish, also follows the "antieli-tist" line proclaimed by Nordhaus. "Concerning the rate of pure time prefer-ence, *Stern* follows a decidedly minority paternalistic view ... that ... selects the lowest conceivable value ... according to the a priori philosophical princi-ple of treating all generations equally – irrespective of preferences for present over future utility that people seem to exhibit in their everyday savings and investment behavior."[39] Here Weitzman cannot excise himself from the grips of the psychological fallacy. It is one thing for an individual to employ a pure time preference with respect to one's own well-being; we know what the author of the story of Esau and Jacob thought about such imprudence. It is quite another thing for public policy to value the benefits to people differently sim-ply because they exist at distinct times. In the absence of a compelling moral reason, public policy that differentially values benefits to persons expresses not imprudence, but invidious discrimination. And widespread imprudence provides no justification for discrimination.

Those economists who assert that there should be a positive rate of pure time preference are committed to the claim that costs and benefits to future persons should be discounted simply because they are to future persons. But they are hard pressed to provide a compelling moral reason for such discrimination. Sometimes they try to offer such reasons; other times, some of the same writers inconsistently assert that this is not a moral matter at all. The social discount rate is simply the rate by which we in fact do discount the future benefits of investments. But that tells nothing about what we *should* do. One cannot prescribe policy without establishing any normative claims. That would be like a parent telling a child how much he should bathe simply by citing the frequency of his baths.

It would make far more sense to accept social discount rate for climate change as a moral matter. But then the *moral* problems with the elasticity of marginal consumption and the rate of pure time preference make a difference. These problems go to the heart of the moral assessment of the project of optimizing intergenerational consumption. The conclusion to draw, then, is that another approach to the intergenerational assignment of the costs of climate change is in order.

---

[39] Weitzman, "A Review," 707.

At least within the framework of optimizing intergenerational consumption, however, there is one plausible reason for discounting climate change-caused costs. This has to with the monetary value of goods over time in a growing economy.

ECONOMIC GROWTH

Any social savings policy is a transfer of resources from present people to future people (some of whom may be the same people). Let's imagine a simple case of only two nonoverlapping generations, each with its own income stream (both the populations and the income streams are equal across generations), and a future cost. The cost is foreseeable. So, present people can pay some or all of it now by means of deferring consumption and saving to augment the income of future people. Alternatively, present people can elect to defer all of the costs to future people. One approach that present people might take is to employ a principle that requires intergenerational impartiality of the kind discussed in the previous section. To adopt such a principle is to assume a moral attitude because, absent moral reasons, self-interest would support partiality for one's own projects. Assuming intergenerational impartiality, a transfer of resources from the present to the future to pay the foreseeable cost does not raise any special moral concerns related to time.

Take a particular example of a foreseeable cost. The first generation knows that the second will have to replace some of its public housing stock. The dwellings will cost $100,000 each. There will be a need for ten of them – a bill of $1,000,000. There is no economic growth and no interest on investments. There may be moral reasons that have nothing to do with time that would recommend either paying the cost of the housing now or later. That is not our concern right now. Our question is whether – when comparing the present cost to the future cost, perhaps for the purpose of deciding what morality requires – the two costs should be weighed differently. Money spent on future people is money not available to spend on the current generation. Insofar as the people of the current generation judge impartially, they would not take the expenditure on future people to be any less important than an expenditure of the same amount on the present generation. The morally right choice – whatever standard is used to determine that – would not be affected by the fact that spending the money on the housing stock would benefit people in the future. And, insofar as the current people are not morally required to spend on future people, they are at liberty to spend on themselves.

Now alter the example by supposing that between the two generations the economy, including housing stock, is expected to grow by 5 percent. The number of dwellings needing to be replaced does not change, but because of growth, there are more of these available in the future than now. Demand does not change, but supply increases. John Broome offers a lucid supply-side defense for discounting the present price of future costs in cases of growth like this.[40] People in the present should apply a discount rate when calculating the costs of the dwellings because there are more dwellings available on the market in the future and the actual market price will be sensitive to the change in supply. To compare the price for each generation, given the actual market price for the future dwellings, we must not compare present market price against the future one; rather, we need to discount the present market price. Failure to do so would be to compare a cost in the present to one in the future that is cheaper. Discounting for growth (or the fecundity of technology), in other words, is necessary in order to compare costs properly.

Broome does not employ the supply-side analysis to make a point about its moral import, but there is a moral point to be made. There are two distinct steps in the moral judgment about which generation should assume certain costs, the counting of the costs, and the application of a moral principle for their distribution. At the second step, the utilitarian would apply the principle of optimizing happiness. However, cost counting – not optimization – is my present interest. Discounting for growth is necessary at the first step to get the right sum of costs. But it is noteworthy that even this prior step invokes moral considerations. When Bentham claimed that the utility of each counts for only one, he was not defending the principle of happiness maximization. Rather, he was expressing a counting principle applicable to an earlier step. When tallying up happiness for purposes of comparing the moral merits of different courses of action (which eventually the utilitarian judges according to happiness maximization), we should be impartial with respect to persons. That is a counting principle, but it is a morally significant one. The principle that the present value of future costs should be discounted for growth is analogous. Unless we discount for growth, we will compare the higher present price of future costs to the lower future price of those costs. In the housing example, discounting the present price of the future costs of the houses renders equal the amount that the two generations would pay to replace the housing stock. Discounting for economic growth is morally significant because it preserves intergenerational impartiality in comparing costs. Unlike discounting on the basis of a pure time preference or on the basis of the elasticity of marginal

---

[40] See Broome, *Ethics out of Economics*, pp. 44–67. A demand-side approach is defended by Cline in *The Economics of Global Warming*, pp. 256–257.

consumption, the moral significance of discounting for growth in discounted utilitarianism is not as a reason for consumption, but a means to get the accounting morally right.

One might object that my argument is mistaken about the moral reason for discounting for growth. It's not that a failure to discount is contrary to impartiality; rather, a failure to discount is morally wrong because it necessarily leads to inefficient outcomes and any plausible moral principle will at least prima facie reject inefficiencies (and utilitarianism rejects them categorically). In response to this objection, it is false that a failure to discount for growth *necessarily* leads to inefficient outcomes. If, as in the aforementioned example, the present generation does not at the first step discount the present value of future dwellings in their decision, they will weigh paying more now – in excess of the future market price – for the houses against paying the lower market price later. If they are intergenerational utilitarians, at the second step they would be required to have future generations pay (at the lower price) in order to minimize costs. In other words, a failure to discount for growth would require the utilitarian to assign the costs to the later generation, not to make an inefficient choice.

Consider an analogous case. Suppose a classical utilitarian fails to observe Bentham's principle of impartiality and puts less weight on the happiness of members of group A than on members of B. As a result, after aggregating the interests of the two groups, the utilitarian falsely believes that happiness is to be promoted by helping group B rather than A, when in fact if the interests were counted equally, the utilitarian would be indifferent about whether help members of group A or B. The utilitarian is not acting inefficiently in helping B, but it is nonetheless false that happiness maximization requires that B be helped at the expense of A.

To the present objection then, a failure of impartiality does not necessarily lead to inefficiencies and cannot be proscribed for that reason. If we assume economic growth, there may be a legitimate moral reason for discounting the present value of future costs at the rate of growth. To do so preserves intergenerational impartiality.

This analysis of discounting for growth reveals a devastating moral inconsistency that can arise in the application of the social discount rate, if positive value is assigned to the rate of pure time preference. The reason to discount for growth is to maintain impartiality in counting the costs, but impartiality is violated by a pure time preference for the future, which counts some costs as less important than others for no other reason than that they are borne by people in the distant future. There is, then, a basic moral contradiction in any application of the social discount rate that includes a positive value both to growth and the rate of pure time preference. The moral reason for the former undercuts the latter.

There is a reason in the case of climate change to be less sanguine about discounting for growth. The uncertainty surrounding the severity of climate change should weaken our confidence that future global economic growth will resemble past growth, and it encourages us to consider the possibilities of regional climate change–induced negative growth.[41] It is not obvious over the course of 100 years or more that the present value of future climate change–induced costs can reasonably be discounted for growth on the order of growth over, say, the twentieth century. And looking only at the growth in global consumption may obscure considerable regional variation in growth owing both to the differential regional impact of climate change and to the differential capacities of people to adapt. Discounting sometimes provides a reason for consumption by lowering the assessment of future costs. Insofar as we lack confidence in uniform economic growth, the reason for consumption supplied by discounting for growth is weakened.

HUMAN DIGNITY AND OPTIMIZATION

In Chapter 1 I argued that respecting each person as a possessor of human dignity requires guiding policies by moral principles that are acceptable to everyone seeking agreement that is based on the respect for human dignity. As we have seen, there are serious moral problems with two of the factors of the social discount rate employed by discounted utilitarianism. In addition to those considerations, it is doubtful that the goal of aggregating consumption pursued by discounted utilitarianism is consistent with the requirement on policy-guiding principles imposed by respect for human dignity. As we saw in the discussion of Nordhaus's climate-wrinkle experiment, aggregation of consumption requires a generation to take on costs if doing so would reduce greater costs to a plurality of future generations. By applying a social discount rate, Nordhaus and other discounted utilitarians seek to reduce the especially heavy burden that this can place on a generation. But doing so merely applies a bad fix to a rotten theory.

The application of a policy-guiding principle directed toward increasing aggregate benefits could require a person to carry a burden because doing

---

[41] See also Weitzman, "A Review," 715–723, and Dasgupta, "Discounting," secs. 5–6; and Partha Dasgupta et al., "Intergenerational Equity Social Discount Rates, and Global Warming" in Portney and Weyant (eds.), *Discounting and Intergenerational Equity*, pp. 51–77.

so would make it unnecessary to impose individually smaller burdens of a less-significant kind on others if, in aggregate, these burdens were more. No such principle allowing policy to impose burdens in that manner could be accepted by people viewing themselves as coauthors of common policy based on the respect for human dignity because it would allow placing crushing burdens on a person in order to produce very small benefits on a sufficient number of people. It would be permissible to intern a person in a large gerbil wheel and to make him run in order to generate electricity if doing so allowed a sufficient number of other people to set their air conditioning one degree cooler in the summer. Even if in times of outrage you are tempted to believe that some people deserve the gerbil wheel, it could not be because it would make a sufficient number of others slightly more comfortable.

In this case, what is true of individuals is also true of generations. The only change is that it would be groups of people whose subjugation would be putatively justified in this way. Crushing burdens could be imposed on a large number of people so that many more might individually enjoy a morally trivial benefit. The underlying problem with discounted utilitarianism is that the aggregation that it employs is inconsistent with respecting the dignity of persons.

As a theoretical project, discounted utilitarianism is in moral shambles. A more promising model of intergenerational cost sharing is, I believe, the principle of equalizing the generational costs to benefits ratios.[42] For purposes of guiding policy, however, it may be enough to rely on the importance of precaution.

PRECAUTION AND MITIGATION

The classical utilitarian takes our basic moral duty to be the promotion of maximal happiness. In economics the discounted utilitarian guides policy toward optimal intergenerational consumption. According to this view we should reject a climate change policy (or a lack of policy such as business-as-usual) if it produces an intergenerational growth path that is suboptimal. But as I have been arguing, optimal consumption is not a goal worthy of our moral allegiance; economic growth could only plausibly be good instrumentally. And there is no reason to suppose that the classical utilitarian would endorse

---

[42] For a philosophical defense of this view of intergenerational cost sharing see Darrel Moellendorf, "Justice and the Assignment of the Intergeneartional Costs of Climate Change," *Journal of Social Philosophy*, 40 (2009c): 204–224. An initial attempt to model the idea is discussed in Axel Schaffer and Darrel Moellendorf, "Beyond Discounted Utilitarianism – Just Distribution of Climate Costs," *Karlsruhe Beiträge zur Wirtschaftspolitischen Forschung* 33 (2013) forthcoming.

ceaseless growth in economic output as conducive to happiness. John Stuart Mill certainly did not.

The antipoverty principle identifies a different reason for rejecting business-as-usual. Business-as-usual is identified as dangerous if there is a mitigation policy with fewer poverty-prolonging costs. The defense of this principle that I offered in Chapter 1 was based on two main ideas. One is that respect for human dignity rules out policy-guiding principles that would be unacceptable to people seeking agreement on the basis of respect for human dignity. The other is that involuntary poverty is something that everyone has good reason to avoid. This suggests that it would be a serious wrong to impose poverty-prolonging climate change on people in the future if there is a mitigation plan at comparatively minor costs.

It is to his credit that when Stern seeks to justify a policy directed toward limiting atmospheric stocks of greenhouse gas at a particular ceiling his argument rests on considerations of costs, benefits, and precaution rather than the alleged good of an optimal growth path. He focuses on limiting concentrations of all greenhouse gases to the warming equivalent of 500 ppm of CO_2 (or 500 ppm CO_2e). (The measure of CO_2e includes the warming potential of all greenhouse gases by converting that potential to the common currency of the warming potential of CO_2; it is therefore a more comprehensive measure, but used primarily within scientific discussions.) Given the current consensus regarding climate sensitivity, this is likely to produce an equilibrium warming of about 2.4–2.8°C, and has a 95 percent probability of exceeding the internationally recognized target of 2°C. The argument that Stern makes for this limit of greenhouse gas concentrations is that the costs of reducing emissions so as to not exceed the limit are relatively low – about 2 percent of the global GDP per annum.[43] According to Stern's calculations, the result of this annual cost would be merely that "the world economy would take an additional six months in reaching the level of world income it would otherwise reach by 2050."[44] He compares this modest price to "the risk of genuinely disastrous outcomes for the planet."[45]

Stern's point is basically precautionary, but to make it he maintains that "[s]uch a payment is not very different from the premium to insure against a small probability of a disastrous outcome."[46] This is an instance of the method

[43] Nicholas Stern, The Global Deal: Climate Change and the Creation of a New Era of Progress and Prosperity (New York: Public Affairs, 2009), chp. 3.
[44] Stern, The Global Deal, p. 54.
[45] Ibid., p. 54.
[46] Ibid.

of dealing with uncertainty as if it were a low probability, which I criticized in Chapter 3.[47] Various truly disastrous outcomes, including massive ice sheet collapse, huge methane releases from thawing tundra and warming oceans, widespread food insecurity, and massive species extinction are not estimated to be low-probability events by scientists. Rather, they are matters of cascading uncertainty. For the most part, Stern appreciates this and does not make the case for mitigation by actually comparing the costs of mitigation against the costs of these outcomes discounted by their probability. As a result, his overall argument is compelling.

As discussed in Chapter 3, given the following circumstances, minimax reasoning about possible climate disasters is reasonable: We have very good reasons to avoid the worst outcomes; little reason to care much about the comparatively minimal costs of reducing the probability of their occurrence; although all of the outcomes are epistemically uncertain, they are more or less explicable given the current state of the science; and we know that circumstances are conspiring to produce some of the known causal factors of the worst outcomes. We are justified, then, in taking a precautionary approach to the disasters. Precaution adds to the reasons that future persons would have for avoiding the risks of climate perturbations. Put differently, assuming a plan for a just distribution of the costs of mitigation, precaution supplements the reasons for mitigation.

Stern's account may not be precautionary enough, however. He sets a target for greenhouse gas concentrations that would very likely produce an equilibrium average temperature in excess of the internationally recognized 2°C limit. According to the summary of the scientific literature provided by the IPCC, to have a good chance of limiting warming to 2°C, atmospheric concentrations need to be 350–400 ppm CO_2 (or 445–490 ppm CO_2e).[48] Recall the litany of disasters listed in Chapter 1. A 1°C temperature increase would put up to 30 percent of existing animal and plant life at increased risk of extinction and would increase coral bleaching; a 2°C increase would result in increased damage from floods and storms and in most coral being bleached. Beyond that, matters turn especially grim. A 3°C increase would expose hundreds of millions of people to increased water stress; a 3.25°C increase would bring about increased morbidity and mortality from heat waves, floods, and droughts;

---

[47] John Broome also uses this method in his *Climate Matters: Ethics in a Warming World* (New York and London: W.W. Norton and Co., 2012), p. 131.

[48] Intergovernmental Panel on Climate Change, *Climate Change 2007: Synthesis Report Summary for Policy Makers*, sec. 5.4. http://www.ipcc.ch/publications_and_data/ar4/syr/en/mains5-4.html (accessed November 5, 2012).

and a 3.75°C would cause a 30 percent loss of coastlines and the suffering of millions of people because of flooding.[49]

Stern does not directly argue against this lower concentration target, but instead rejects one even lower still – namely, 400 ppm CO_2e – which is approximately 350 ppm CO_2. Stern's argument against this target is that because it is already below existing atmospheric concentrations of CO_2, achieving it would require immediate reductions in emissions so drastic that the result would be a prolonged global economic recession.[50] The global poor would suffer tremendously if that turned out to be the case. If Stern's projections are correct, then there is unlikely to be a mitigation plan in accordance with the lower limit on CO_2 concentrations that can satisfy the antipoverty principle defended in Chapter 1:

> Policies and institutions should not impose any costs of climate change or climate change policy (such as mitigation and adaptation) on the global poor, of the present or future generations, when those costs make the prospects for poverty eradication worse than they would be absent them, if there are alternative policies that would prevent the poor from assuming those costs.

Because both current and future people have reasons deriving from the goal of eradicating poverty to consume fossil fuel, there must be a strong presumption against any mitigation plan that cannot satisfy the antipoverty principle. Still, Stern's argument against the lower concentration target of 400 ppm CO_2e or 350 ppm CO_2 does not directly address the higher range of 445–490 ppm CO_2e thought to be reasonably likely to limit warming to 2°C. If this target could be hit without causing a global recession, then Stern's argument would not apply to it. He has not provided an argument to diverge from the internationally agreed upon warming limit of 2°C limit.

AN EMISSIONS BUDGET

If we assume that precaution supports limiting warming to 2°C, following the important work of the philosopher Henry Shue, our mitigation duties to future generations can be reasonably clearly specified in terms of an emissions budget.[51] A growing body of scientific research supports the

[49] *Ibid.*, 3.3.1. http://www.ipcc.ch/publications_and_data/ar4/syr/en/mains3-3-1.html (accessed November 5, 2012).

[50] Stern, *The Global Deal*, p. 39.

[51] See Henry Shue, "After You: May Action by the Rich Be Contingent Upon Action by the Poor?" *Indiana Journal of Global Legal Studies* 1 (1994): 352–353; Henry Shue, "Historical Responsibility," Technical Briefing for Ad Hoc Working Group on Long-term Cooperative Action under

claim that achieving 2°C limit requires limiting cumulative carbon emissions to one trillion tons.[52] Keeping within the budget requires a transition to a no-carbon economy. Because constraining emissions on precautionary grounds is for the sake of future generations, I call the budget *humanity's morally constrained CO$_2$ emissions budget*. The constraint is also morally significant now and in the near future because presently the cheapest fuel for propelling human development and liberating people from poverty is coal.

From the dawn of the Industrial Revolution until now, the cumulative carbon emissions have been about a half a trillion tons. At the current business-as-usual emissions level the rest of the morally constrained budget will be used up in 2041 or sooner.[53] In other words, we are on course to consume the other half of the budget within the lifetime of many people alive today. If we use up the budget entirely and fail to provide subsequent generations with an inexpensive means to generate energy, we will have done a serious injustice to those who follow us.

A morally constrained budget is different than one in which the constraints are physical or financial. If we use all the water in the well, we go thirsty. If we spend our earnings before the next payday, we go without money. If we emit the next half-trillion tons, we do not prevent people after 2041 from emitting; rather, we leave them with the choice of either continuing to emit and thereby imposing risks and uncertainties on future generations that exceed precaution or using more expensive renewable energy and thus delaying human development. To leave the people with such a desperate choice would be

---

the Convention [AWG-LAC], SBSTA, UNFCC, Bonn, June 4, 2009, 4–7. http://unfccc.int/files/meetings/ad_hoc_working_groups/lca/application/pdf/1_shue_rev.pdf (accessed June 27, 2013); and Henry Shue, "Climate Hope: Implementing the Exit Strategy," *Chicago Journal of International Law* 13 (2012): 394–395.

52 See Malte Meinshausen, et al., "Greenhouse-gas emission targets for limiting global warming to 2°C," *Nature* 458 (2009): 1158–1162; Myles R. Allen, et al., "Warming Caused by Cumulative Carbon Emissions towards the Trillionth Tonne," *Nature* 458 (2009): 1163–1166; H. Damon Mathews, et al., "The Proportionality of Global Warming to Cumulative Carbon Emissions," *Nature* 459 (2009): 829–832; H. Damon Matthews, et al., "Cumulative Carbon as a Policy Framework for Achieving Climate Stabilization," *Philosophical Transactions of the Royal Society A* 370 (2012): 4365–4379. According to Intergovernmental Panel on Climate Change, *Climate Change 2013, The Physical Science Basis, Summary for Policymakers*, 25, capping cumulative carbon emissions at 1 trillion tons establishes a greater than 66% chance of limiting warming to 2°C. This, however, does not include the radiative forcing caused by other greenhouse gases. When they are included, cumulative carbon emissions should cap at 790 GtC (billion tons of carbon).

53 See the website http://trillionthtonne.org/ (accessed July 4, 2013).

deeply disrespectful of their capacity for moral choice.[54] It also would implicate us either indirectly in the suffering of more-distant generations or directly in the suffering, owing to prolonged underdevelopment, of the generation of 2041.

[54] For a discussion of placing future generations in a worse moral position, see also Henry Shue, "Responsibility to Future Generations and the Technological Transition," in Walter Sinnot-Armstrong and Richard B. Howarth, eds. (Amsterdam and San Diego: Elsevier, 2005), pp. 272–275.

5

The Right to Sustainable Development

"Those Greek were superficial – *out of profundity*."
– Friedrich Nietszche

The seventeenth Conference of the Parties of the United Nations Framework Convention on Climate Change (UNFCCC) – meeting in Durban, South Africa, in 2011 – agreed on the Durban Platform for Enhanced Action, which contains a commitment "to develop a protocol, another legal instrument or an agreed outcome with legal force under the Convention applicable to all Parties."[1] The negotiations around such a protocol will be conducted within the framework of norms expressed in the Convention. Analogous to a game, these norms are the rules according to which the parties to the Convention have agreed to play when deliberating about joint action and supplemental treaties, such as the one they committed to develop in the Durban Platform. If taken seriously, the norms would play an important role in constraining negotiations, thereby narrowing the range of disagreement.

This chapter discusses one especially important norm for the morality of climate change policy expressed in the Convention; the right to sustainable development is stated in Article 3, paragraph 4:

The Parties have a right to, and should, promote sustainable development. Policies and measures to protect the climate system against human-induced change should be appropriate for the specific conditions of each Party and should be integrated with national development programmes, taking into

---

[1] United Nations Framework Convention on Climate Change, *Report of the Conference of the Parties on Its Seventeenth Session, Held in Durban from 28 November to 11 December 2011*. http://unfccc.int/resource/docs/2011/cop17/eng/09a01.pdf#page=2 (accessed November 22, 2012).

account that economic development is essential for adopting measures to address climate change.[2]

Sustainable development is understood in many different ways. But in the context of the Convention and other related UN resolutions, its meaning is fairly determinate. I call the conception of the right to sustainable development more or less fixed by this context *the institutional conception of the right to sustainable development* in order to distinguish it from the many other understandings of sustainable development in use.

The fact that these various states – but most importantly the United States, Canada, the countries of the European Union, Australia, and Japan – have agreed to constrain proposals for avoiding dangerous climate change by a norm that recognizes the right to sustainable development constrains acceptable mitigation and adaptation proposals in a morally salutary manner. Moreover, the international pursuit of climate change mitigation is constrained by fair terms of cooperation, deriving from the very concerns that make mitigation important. In this chapter I argue that these considerations justify the claim that states have the right to sustainable development. The right to sustainable development is a claim that least-developed and many developing states have the liberty to pursue energy-intensive development and not be yoked with the same financial burden to mitigate that highly developed states must wear. This is a freedom not to have to assume an equal share of the demands of intergenerational justice, placed on the present generation on behalf of future generations on precautionary grounds.

A key feature of the justification of this moral claim is that it is not based primarily on a more comprehensive, and more controversial, account of global justice. Instead, there are two defenses. One employs the relatively noncontroversial idea of the reasonableness of acting in good faith in deliberation. The other is derived from the value of fair terms of cooperation in pursuing the international goal of climate change mitigation. Insofar as the justification of the right to sustainable development is not premised on a fundamental good or a comprehensive account of global justice, it is, in that sense, superficial. But because it is based on a noncontroversial deliberative norm, it is also compelling. Sometimes, it seems, the superficial can be compelling.

I begin by elucidating the content of the institutional conception of the right to sustainable development. I then discuss the constraints that this conception of the right places on reasonable proposals in the UNFCCC deliberative

---

[2] United Nations Framework Convention on Climate Change (UNFCCC), Art. 3, para. 4. http://unfccc.int/files/essential_background/background_publications_htmlpdf/application/pdf/conveng.pdf (accessed November 1, 2012).

context. Following that, I sketch the idea of the reasonableness of keeping prior commitments. I then discuss the kinds of circumstances that would render an international commitment illegitimate and argue that these do not apply in the present case. This justification of the reason to respect the right to sustainable development is analogous to the reasons that individuals have to honor promises they make. Normally such reasons are independent of whether it was good to have made the promise. This is a characteristic feature of the superficial justification. I discuss additional reasons to take the right to sustainable development seriously. These draw on the value of fair terms of cooperation in an international framework for climate change mitigation and rely on the concerns of the antipoverty principle.

THE INSTITUTIONAL CONCEPTION OF THE RIGHT

In the context of the United Nations, the concept of sustainable development has a fairly determinate content. United Nations General Assembly resolution 42/187 of 1987, which welcomed the Brundtland Report, "Our Common Future," affirms "that sustainable development, which implies meeting the needs of the present without compromising the ability of future generations to meet their own needs, should become a central guiding principle of the United Nations, Governments and private institutions, organizations and enterprises."[3] This passage is drawn verbatim from the Brundtland Report, chapter 2, paragraph 1.[4]

The Brundtland Report makes an important distinction between sustainable development and economic growth:

> [S]ustainable development clearly requires economic growth in places where such needs are not being met. Elsewhere, it can be consistent with economic growth, provided the content of growth reflects the broad principles of sustainability and non-exploitation of others. But growth by itself is not enough. High levels of productive activity and widespread poverty can coexist, and can endanger the environment.[5]

Human needs are manifold, but the Brundtland Report speaks of "essential needs . . . for food, clothing, shelter, [and] jobs."[6] This list of essential needs

---

[3] United Nations General Assembly, A/RES/42/187, 96th plenary meeting, December 11, 1987. http://www.un.org/documents/ga/res/42/ares42-187.htm (accessed November 1, 2012).
[4] Report of the World Commission on Environment and Development, *Our Common Future*, chp. 2, para. 1. http://www.un-documents.net/wced-ocf.htm (accessed November 1, 2012).
[5] *Ibid.*, chp. 2, para. 6.
[6] *Ibid.*, chp. 2, para. 4.

might be understood to rest on a conception of human nature, but perhaps it is not most charitably read that way. We do not need jobs in virtue of our nature. And although we need nutrition and shelter naturally, food sometimes has other, more socially rich connotations. This list – or at least some of the items on it – could also be understood to refer to the basic needs of typical persons in modern societies.[7] Elsewhere the Brundtland Report speaks more expansively of "human needs and aspirations."[8]

In the Convention, however, the right to sustainable development is not limited to meeting basic needs. Rather, it licenses energy-intensive and poverty-eradicating economic growth. In its preamble, the Convention claims to be "taking into full account the legitimate priority needs of developing countries for the achievement of sustained economic growth and the eradication of poverty," and it recognizes "that all countries, especially developing countries, need access to resources required to achieve sustainable social and economic development and that, in order for developing countries to progress towards that goal, their energy consumption will need to grow."[9] The paragraph that follows the declaration of the right holds that "[t]he Parties should cooperate to promote a supportive and open international economic system that would lead to sustainable economic growth and development in all Parties, particularly developing country Parties, thus enabling them better to address the problems of climate change."[10]

In the Brundtland Report, the qualifier "sustainable" expresses a limitation on development in order to maintain intergenerationally the human capacity to satisfy needs. Sustainability here contains a core notion of fairness to other (future) persons, not first and foremost natural preservation. The Brundtland Report is explicit about this: "Every ecosystem everywhere cannot be preserved intact. A forest may be depleted in one part of a watershed and extended elsewhere, which is not a bad thing if the exploitation

---

[7] Charles Beitz, *The Idea of Human Rights* (Oxford: Oxford University Press, 2009), p. 110. Beitz argues that human rights are "important in a wide range of typical lives that occur in contemporary societies." I am not here invoking Beitz's account of human rights, but merely adopting a similar approach to interpret how needs are understood in the UN context. There is, of course, a rich philosophical literature on human rights. James Griffin justifies them on the basis that they protect personhood. See his *On Human Rights* (Oxford: Oxford University Press, 2008), pp. 35–36. Rainer Forst defends them on the basis of a fundamental right to justification in his "The Justification of Human Rights and the Basic Right to Justification: A Reflexive Approach," *Ethics* 120 (2010): 711–740. And James W. Nickel makes a pluralist defense of human rights in *Making Sense of Human Rights*, 2nd ed. (Malden: Blackwell Publishing, 2007), pp. 53–64.

[8] World Commission on Environment and Development, chp. 2, para. 15.

[9] UNFCCC, preamble.

[10] UNFCCC, Art. 3, para. 5.

has been planned and the effects on soil erosion rates, water regimes, and genetic losses have been taken into account."[11] Preserving the capacity of future generations to meet their needs may require the conservation of some natural resources, but also the transformation of others into capital assets. This anthropocentric conception of sustainability is echoed in Principle 1 of the Rio Declaration: "Human beings are at the centre of concerns for sustainable development."[12]

In the Convention, the bearers of the right to sustainable development are taken to be Parties to the Convention or states. Facially the right is a group right because it is possessed and exercised by states rather than individuals.[13] Persons do not develop in the relevant sense; only states do. So, the right protects the capacity of states, not persons. This, of course, does not entail that it has nothing to do with the moral claims of persons, as I shall discuss later in this chapter. This affirmation of a group right is consistent with other forms of treaty law under the auspices of the United Nations. Article 1, paragraphs 1 and 2 of both the International Covenant on Civil and Political Rights and the International Covenant on Economic, Social, and Cultural Rights declare the rights of people to self-determination and to the disposal of natural resources. The right to sustainable development fits with these sweeping expressions of state sovereignty insofar as it attributes to a state's license over the domain of development-conducive macroeconomic policy. I do not suppose in general that sovereignty should be protected by such sweeping expressions.[14] In fact, I shall argue that in the case of climate change and climate change policy, justified state policies will have to take into account their effects on persons

---

[11] *Ibid.*, chp. 2, para. 11.

[12] United Nations Environment Programme, "Rio Declaration on Environment and Development," Principle 1. http://www.unep.org/Documents.Multilingual/Default.asp?documentid=78&articleid=1163 (accessed November 5, 2012) In Principle 3, the right to development has no qualifier "sustainable," although the principle requires meeting the needs of future generations.

[13] According to one account, group rights "reflect a certain conception of community life that can only be implemented by the combined efforts of every one." See Karl Vasak, "A 30-Year Struggle: The Sustained Effort to Give Force of Law to the Universal Declaration on Human Rights," *UNESCO Courier* 30 (1977): 29.http://unesdoc.unesco.org/images/0007/000748/074816eo.pdf#48063 (accessed December 27, 2012). According to Beitz, they involve "values whose importance for the individuals who enjoy them can only be explained by referring to the facts of these individuals' group membership." See Beitz, *The Idea of Human Rights*, p. 113. But in the Convention, sustainable development is a group right in the sense that only societies can possess the property of being developed.

[14] For sovereignty as final authority in a domain, see Darrel Moellendorf, *Cosmopolitan Justice* (Boulder: Westview Press, 2002), p. 103. I argue for limitations on state sovereignty in *Cosmopolitan Justice*, pp. 104–105.

around the world. And at least in that sense, states are not at moral liberty to pursue whatever energy policies they choose.[15]

DEVELOPMENT

How should we understand claims such as the following two: (1) Country X is more developed than Country Y or (2) Country Z has developed significantly in the last twenty years? The concept of development involves change, but not only change; the meaning of the two claims would not be captured by stating that Country X is more changed than Country Y or that Country Z has changed significantly in the last twenty years. The concept of development includes not only change, but change in a positive direction. In this sense, societal development is akin to biological development (but not in the sense that it is natural). A child's development is not merely measured by physical change, but also by intellectual growth in an anticipated and desired direction. The concept of societal development is inescapably moralized.

One sometimes hears critics doubting whether development is good.[16] Given the necessarily moral character of the concept development, moral criticism of it might seem senseless. It would seem that the only true antidevelopment position is an anti-all-change position.[17] Perhaps there are conservative defenders of the status quo, with all of its poverty and hierarchical traditional relations, but the critics of development are most charitably understood as challenging only the merits of certain conceptions of development.[18] Often these critics rightly challenge conceptions of development as measured only in terms of economic growth. But conceptions of development vary according to the values used to give the appropriate direction of societal change. And there are much more morally compelling conceptions of development than those based only on the value of economic growth.

There is nothing desirable about economic growth for its own sake. It is only important to the extent that it serves other important goods. A more

---

[15] This case is made well in Henry Shue, "Eroding Sovereignty: The Advance of Principle," in Robert McKim and Jeff McMahan, eds., *The Morality of Nationalism* (Oxford: Oxford University Press, 1997), 340–359.

[16] An influential volume of post-development critique is Wolfang Sachs ed., *The Development Dictionary*, 2nd ed. (London: Zed Books, 2010). See, for example, the essay by Gustavo Esteva, "Development," pp. 1–23.

[17] Genuine anti-development positions are rare. Plato has Socrates expressing one by criticizing politicians who giving Athenians "their full of what they desired" made the city "festering and rotten to the core . . . glutted . . . with harbours and dockyards and walls and tributes and rubbish of that sort . . . " Plato, *Gorgias* (518e–519a) Walter Hamilton, trans. (London: Penguin Classics, 2004), p. 124.

[18] I owe this distinction between a concept of something and the conception of it to John Rawls. See his *A Theory of Justice*, rev. ed. (Cambridge, MA: Harvard University Press, 1999), p. 9.

morally compelling conception of development is that of human development as employed by the United Nations Human Development Programme (UNDP). The conception of human development provides a cardinal measure of average human well-being within states and an ordinal ranking of the development levels of states. The constituents of human well-being are taken to be average life expectancy, average educational attainments (understood as a weighted combination of adult literacy rate and the percentage of the school-aged population enrolled in school), and per capita income. These three indices are averaged to obtain a cardinal number between zero and one and an ordinal list of rankings on the Human Development Index. For example, in the UNDP's *Human Development Report of 2009*, Norway had the highest cardinal measure of human development with a score of 0.900. In comparison, the United States, which had a cardinal measure of 0.894, is in thirteenth place ordinally out 182 measured states.[19] (I use UNDP data from 2009 in this discussion because the emission data available to compare it with is from 2009. The Human Development Index did not distinguish between high and very high human development in this report as it did in subsequent years.)

The economist Mahbub ul Haq and others developed this particular conception of development as an alternative to measures and rankings that consider only per capita income. Measuring development is not like measuring the mass of objects. The appropriateness of the Human Development measure depends primarily on the moral importance of the specific indices used and the usefulness of a measure that provides aggregate data for states. Depending on one's purposes, there may be other, better measures of development. The UNDP uses other measures for other purposes; for example, it uses a Gender Development Index to measure human development in a manner that is sensitive to gender inequalities in life expectancy, educational attainment, and per capita income.[20]

Amartya Sen, the 1998 Nobel Laureate in Economics, argues that human development is morally important because improvements in health and education and increases in income expand areas of human freedom.[21] But even if improvements in health and education and increases in income do expand our freedom, the value of these goods does not seem limited to their service in promoting freedom. Improvements in health prolong lives and often relieve

---

[19] United Nations Human Development Programme, *Human Development Report 2009, Overcoming Barriers: Human Mobility and Development*, p. 171. http://hdr.undp.org/en/media/HDR_2009_EN_Complete.pdf (accessed November 1, 2012).

[20] United Nations Development Programme, *Human Development Report 2007–2008, Fighting Climate Change Human Solidarity in a Divide World*, "Technical Note 1," pp. 355–356. http://hdr.undp.org/en/media/HDR_20072008_Tech_Note_1.pdf (accessed November 1, 2012).

[21] Amartya Sen, *Development as Freedom* (New York: Anchor, 2000).

suffering. And education is also important to active citizenship and in acquiring a decent job. The expansion of human knowledge, which education can facilitate, is perhaps valuable for its own sake. The pursuit of basic research in physics and mathematics has an instrumental value, but it may be valuable simply as a means of expanding our knowledge. And increased income is valuable for several reasons: It can lift people out of desperate poverty, it can be useful for attaining health and education, and shelter and security often require money. Income provides the means for the pursuit of many ends that we value. There is a tendency to a monism about value in Sen's account that seems both unnecessary and implausible. The goods of human development are important for many reasons in addition to their role in expanding human freedom.

Justice may be the first virtue of social institutions, as John Rawls famously claims, but it is not the only virtue.[22] A just society in which educational attainments are high and the population is healthy is better than a just society in which everyone is deeply impoverished. Human development is a virtue of societies because its constituents are morally valuable, and for a variety of reasons. Hence, the moral reasons we have for caring about human development are different than the reasons we have to care about justice. This can be captured by a distinction between the production and the distribution of morally valuable goods. We care about development because we care about the production of morally valuable goods. We care about justice (in part, at least) because we care about the distribution of these goods.

The production of the goods that comprise human development is an urgent moral project. Our world is marked by tremendous misery. More than 40 percent of the world's population lives each day on less than the equivalent of what $2 buys in the United States.[23] Easily prevented or curable diseases kill about 20,000 children each day.[24] And opportunities for education are almost nonexistent for children born into these brutal conditions and sent off to work at an early age. This situation is plausibly a global injustice.[25] Part

[22] Rawls, *Theory*, p. 3.
[23] World Bank, "Poverty." http://web.worldbank.org/WBSITE/EXTERNAL/TOPICS/EXTPO VERTY/EXTPA/o,,contentMDK:20040961~menuPK:435040~pagePK:148956~piPK:216618 ~theSitePK:430367~isCURL:Y,oo.html (accessed November 1, 2012).
[24] UNICEF, *State of the World's Children 2012 Children in an Urban World, Executive Summary*, p. 4. http://www.unicef.org/publications/files/SOWC_2012-Executive_Summary_EN_ 13Mar2012.pdf (accessed November 1, 2012).
[25] There are various defenses of this claim. See, for example, the arguments developed in Gillian Brock, *Global Justice: A Cosmopolitan Account* (Oxford: Oxford University Press, 2009); Simon Caney, *Justice Beyond Borders: A Global Political Theory* (Oxford: Oxford University Press, 2006); Darrel Moellendorf, *Global Inequality Matters* (Basingstoke: Palgrave

of the remedy must include protection of, and support for, the capacities of states to pursue human development. As I hope will become clear, securing protection for the capacities of states to pursue human development in the context of an international climate change treaty places a constraint on the energy policies of highly developed states.

SUSTAINABILITY

How does sustainability limit the license of a state to pursue development?[26] One answer is that the development policy of a state should be scrutinized to see if its policy, taken in isolation, is consistent with the norm of sustainability. There is language in the Brundtland Report that suggests this approach: "A society may in many ways compromise its ability to meet the essential needs of its people in the future – by overexploiting resources, for example."[27] This answer might be credible in cases in which the resource is peculiar to the state (for example, aquifers that are not shared by other states), but it is less plausible when the resource is shared with other states. The Brundtland Report seems aware of this and suggests a standard that takes into consideration the policies of other states: "Living standards that go beyond the basic minimum are sustainable only if consumption standards everywhere have regard for long-term sustainability."[28] According to this standard, a state's policies are unsustainable if they produce living standards beyond a basic minimum and if the standards of other states are also unsustainable. This, however, does not provide sufficient clarity concerning the standard of sustainability because it takes unsustainability in one state to be a function of unsustainability in others, and thereby produces a regress problem. But it does suggest something helpful; namely, that sustainability – at least with respect to global natural resources – is a property not merely of the policies of a single state, but of all states taken together.

The natural processes that recycle and reabsorb greenhouse gasses are global resources. Sustainability, then, should be directed to the sum of all greenhouse

---

Macmillan, 2009); and Thomas Pogge, *World Poverty and Human Rights*, 2nd ed. (London: Polity, 2008).

[26] The institutional conception of the right to sustainable development skirts a great deal of philosophical controversy about the norm of sustainability. See Brian Barry, "Sustainability and Intergenerational Justice," *Theoria* 45 (1997): 43–65 and Bryan Norton, "Ecology and Opportunity: Intergenerational Equity and Sustainable Options" in Andrew Dobson, ed., *Fairness and Futurity* (Oxford: Oxford University Press, 1999), pp. 118–150. These discussions are important for understanding the concept of sustainability more generally.

[27] World Commission on Environment and Development, chp. 2, para. 8.

[28] *Ibid.*, chp. 2, para. 5.

gas emissions. It is the sum of all emissions that is presently exceeding the capacity of these processes to recycle CO_2 without climate perturbations. A state may emit more than average, but this is sustainable only so long as the total global emissions are consistent with avoiding dangerous climate change. Sustainability with respect to greenhouse gas emissions is a constraint on total emissions, not on the emissions of a particular state taken in isolation.

International policy does not, then, contravene the constraints of sustainability simply because it allows some states to emit more than would be sustainable if all states emitted the same total or per capita amount. The limitation that sustainability puts on development is violated only if global emissions are unsustainable. The right to sustainable development is the right to develop in the context of a sustainable global energy policy.

SUSTAINABLE DEVELOPMENT AND CLIMATE CHANGE POLICY

Progress in achieving a high level of human development requires energy – a lot of energy. In order to develop rapidly, countries need access to cheap forms of energy. Currently fossil fuels are much cheaper than renewable energy. To prevent the slowing of the human development process, least-developed and some developing countries need to be able to purchase energy at low prices. There are two ways that an international treaty could permit this: (1) Either these countries could be allowed to use fossil fuels without the same restrictions placed on highly developed countries or (2) the purchase of renewable energy and renewable energy-generating technology could be subsidized by highly developed countries.

By making several assumptions, we can see most clearly how the institutional conception of the right to sustainable development constrains acceptable proposals for climate change policy. First, I assume that the only policy lever available for mitigating climate change is the maintenance of a schedule of CO_2 emissions reductions. According to this assumption, the task of respecting the right to sustainable development involves ensuring that least-developed and some developed states are allowed emissions allotments sufficient to achieve development within a plan of global emissions reductions.

Second, I assume that an international agreement should be committed to the goal of limiting warming to no more than 2°C (about 3.4°F). The discussions of Chapters 3 and 4 suggest that there are two good reasons for wanting to keep warming close to this limit. The first is the cascading uncertainty of catastrophic events as global mean temperature increases. The second is the relative economic ease with which the limit could be met. As we shall see, assuming this limit simply makes it easier to illustrate the relationship between limiting greenhouse gas emissions and achieving human development. The point remains wherever the temperature limit is set.

Third, I assume prices for fossil fuel and energy from renewable sources at roughly current levels. These could change significantly if there were a technological breakthrough or an increase in the cost of emitting CO_2. Finally, I assume that the sort of development worth pursuing is the UNDP rank of high human development. The point of all these assumptions is simply to make it easier to see how the institutional conception of the right to sustainable development might constrain acceptable proposals for climate change policy. The actual constraints on policy will vary as conditions change and according to other conceptions of development. But those possibilities would not affect the basic point that the right to sustainable development constrains acceptable proposals within the context of the Convention.

We can see what sort of allowance should be made for human development by examining the emission of highly developed states. How much do most highly developed states currently emit? The only useful measure is per capita CO_2 emissions because counting total emissions does not control for population size and, therefore, does not sufficiently isolate the relationship between development and emissions. There are thirty-eight states in the 2009 United Nations Development Programme's highly developed category, ranging from Norway in first place to Malta in thirty-eighth.[29] The list of the states in between includes the United States in thirteenth place, Singapore in twenty-third, South Korea in twenty-sixth, and Portugal in thirty-fourth. Of these thirty-eight states, Portugal has the lowest per capita CO_2 emissions at 5.4 metric tons (mt). In comparison, Norway's per capita emissions are 8.7 mt, the United States' are 19.18 mt, South Korea's are 11.2 mt, and Malta's are 7.9 mt.[30]

Looking at the per capita emissions of these states, the lesson is unclear. South Korea is the newcomer on this list. In 1975 South Korea had the same human development rank as Jamaica, which in 2009, at ordinal rank 100, is still below the median and solidly in the middle of the medium human development group. We might conclude, then, that rapid human development requires per capita CO_2 emissions of approximately 11 mt. On the other hand, Singapore – up from thirty-fifth place in 1990 – has staggering per capita emissions of 34.6 mt. Perhaps it is, then, too much to expect a state to achieve high human development while emitting only 11 mt of CO_2 per capita. Singapore,

[29] UNDP, 2009, pp. 169–174.
[30] United States Energy Information Administration, "International Energy Statistics." http://www.eia.doe.gov/oiaf/ieo/highlights.html (accessed November 1, 2012).

however, is probably an exception. It relies heavily on entrepôt trade, importing vast amounts of oil, refining it, and re-exporting it. This contributes to its remarkably high level of per capita emissions. Moreover, Portugal is able to *maintain* its presence among the group of states characterized as containing high human development with per capita emissions of 5.4 mt. Still, expecting a state to *achieve* high human development with emissions similar to Portugal's might be unrealistic because it might be easier to invest in less carbon-intensive forms of energy production and use once a state is already highly developed. It would seem a case of misplaced confidence to rely too heavily on any one emissions figure from the survey of the most highly developed countries. Instead, the range of 5–11 mt of CO_2 per capita will serve as a reference for the following discussion of emissions and human development.

The Intergovernmental Panel on Climate Change (IPCC) holds that limiting warming to 2°C will probably require global CO_2 emissions 50–85 percent below 2000 levels by 2050.[31] Average global per capita CO_2 emissions in 2000 were 3.92 mt for a population that was just over 6 billion.[32] If we assume that population growth yields a global population of 9 billion in 2050, average per capita emission entitlement given a 50–85 percent global reduction would be in the range of 0.4–1.33 mt of CO_2. Of the ninety-one countries in the top half of the UNDP's 2009 Human Development Index, only two – Albania and Peru – have per capita emissions within the 0.4–1.33 mt range. The survey of states with high human development suggests that we have good reasons to believe that making significant advances toward high human development will require average per capita emissions to be considerably higher than the 0.4–1.33 mt range, perhaps between 5 and 11 mt of CO_2 per capita.

THE ASSUMPTIONS REVISITED

The previous discussion is based on several assumptions. For one, it assumes that the only policy lever available for climate change mitigation is a schedule of CO_2 emissions reductions. In reality, an international treaty on mitigation could combine constraints on emissions with the requirement that highly developed states subsidize the purchase of renewable energy on the part of least-developed and developing states so that human development can be pursued without cost increases and with lower emissions. But this merely alters

---

[31] Intergovernmental Panel on Climate Change, *Climate Change 2007: Synthesis Report*, sec. 5.4. http://www.ipcc.ch/publications_and_data/ar4/syr/en/mains5-4.html (accessed November 1, 2012).

[32] U.S. Energy Information Administration, "International Energy Statistics."

the action required by the duty to respect the right to sustainable development. The duty could be met by not constraining the capacity of states to emit CO_2; it could also be met by a provision of resources, technological knowledge, or both to least-developed and developing states. Both kinds of action correspond to claims afforded to least-developed and developing states on grounds of their liberty to pursue development within the constraints of global sustainability.

Moreover, an international policy will also have to address adaptation because even limiting warming to 2°C will produce sea-level rise and climatic changes. Some of this will threaten human development gains. The provision of resources to certain developing and least states will be necessary just to sustain human development gains and to continue where progress is being made. The right to sustainable development in the context of formulating adaptation policy is not plausibly limited only to the duty of allowing states to develop within a globally sustainable framework. But once again, this is just to reinforce the main point of this chapter: namely, that the institutional conception of the right to sustainable development places constraints on policies that may be pursued consistently with the aim of limiting global warming and its negative effects.

The previous discussion also assumes that global climate change policy should be consistent with all states attaining the level of human development characterized as "high" by UNDP criteria. As we have seen, this is not how the right to sustainable development is understood in the Brundtland Report. It speaks merely of development as meeting basic needs. If the Brundtland Report's conception were taken as the ideal of development, the right to sustainable development would be less constraining on developed states than the aforementioned discussion suggests. It seems unlikely, however, that a proposal that limits the ambitions of least-developed and developing states merely to meeting basic needs would be considered reasonable within UNFCCC negotiations. As we have seen, the Convention is concerned to license energy-intensive and poverty-alleviating economic growth. Moreover, the goal of climate change negotiations is to avoid dangerous climate change but within an equitable framework for energy use. A proposal that would achieve the aim of protecting the climate system by means of a principle that would restrict the development ambitions of some countries while allowing high human development in countries that have already achieved it would be objectionable on grounds of distributing the burdens of achieving the good of sustainability in an unfair manner.

Finally, the previous discussion makes assumptions regarding the volume of CO_2 emissions necessary to achieve high human development. If there were a technological breakthrough in energy generation or conservation, satisfying the constraint that sustainability puts on the right to development would be less onerous. A fundamental goal of international policy must be to promote such a breakthrough by raising the costs of CO_2 emissions.[33] This would lower the price of renewable energy in comparison to fossil fuels, increasing demand for it and providing incentives for investments in innovation that will lower the absolute price of such fuels.

REASONABLENESS AND GOOD FAITH DELIBERATION

The Convention is the treaty framework in which multilateral negotiations concerning climate change under the UN auspices take place. These Conferences of the Parties (COPs) are the deliberative bodies in which international climate change policy is negotiated, proposed, and decided upon. Hence, the Convention's norms constrain and direct the negotiations of parties and limit the proposals that a party can reasonably make within the Convention. I am using a familiar sense of the term "reasonable," which is applicable, for example, to constraining one's conduct to the terms of a prior agreement. It is reasonable to do as one undertakes to do.

The role of deliberative norms in the present case is to guide further deliberation in an effort to make advances in an unfinished project of treaty making. Proposals that contradict the norms of the deliberative framework undermine the deliberation-guiding role of the framework and threaten the process of deliberation itself. Any party knowingly making such proposals is, in the absence of excusing conditions, negotiating in bad faith and is undermining the agreed upon framework for advancing the project. The actions of such a party appear to be expressing contempt for other parties. An important moral failing, then, of the parties who intentionally ignore the constraints of the Framework, to which they have assented, is that they unreasonably undermine the negotiating process. And parties, in a deliberative process, express

---

[33] This is hardly a novel or controversial claim. Henry Shue has argued for it for years in several seminal essays. See Henry Shue, "Avoidable Necessity: Global Warming, International Fairness, and Alternative Energy," in Ian Shapiro and Judith Wagner DeCew, eds., *NOMOS XXXVII: Theory and Practice* (New York: NYU Press, 1995), p. 259; Henry Shue, "Environmental Change and the Varieties of Justice," in Fen Osler and Judith Reppy, eds., *Earthly Goods: Environmental Change and Social Justice* (Ithaca, NY: Cornell University Press, 1996), p. 27; Henry Shue, "Responsibility to Future Generations and the Technological Transition," in Walter Sinnott-Armstrong and Richard B. Howarth, eds., *Perspectives on Climate Change: Science, Economics, Politics, Ethics* (Amsterdam and San Diego: Elsevier, 2005), p. 273; and Henry Shue, "Climate Hope: Implementing the Exit Strategy," *Chicago Journal of International Law* 13 (2013): 396–398. Richard A. Posner also defends the view in his *Catastrophe: Risk and Response* (Oxford: Oxford University Press, 2004), pp. 156–158.

contempt for their partners when the parties fail to constrain their actions in accordance with the reasonable expectations that they have created.

REASONABLE EXCUSES

There are conditions in which one might be morally excused for neglecting some legal norms previously agreed upon. For example, exceptional or emergency circumstances could require acting in illegal ways to prevent moral disaster. In 1994 when Rwandan Hutu militias slaughtered hundreds of thousands of Tutsi and moderate Hutus and the UN Security Council failed to offer protection to those attacked, it would not necessarily have been unreasonable for a party to act unilaterally in violation of international law to protect those who were threatened. Additionally, morality often requires revising conventional norms, and sometimes those engaging in civil disobedience may be excused for violating legal norms as part of a morally responsible strategy of normative revision.[34] But generally, with respect to a norm that is part of the framework for deliberation for the purposes of further lawmaking, neither of these excusing conditions is likely to exist. With respect to the first excusing condition, emergencies calling for swift action in contravention of a law are more typical than emergencies calling for rapid legislation. Any proposal that might be excused in this way should be explicitly defended as an emergency. With respect to the second, in the deliberative context it is difficult to imagine why a proposal to revise the background norms should not simply be introduced rather than a proposal that contradicts those norms. In the context of norm-ordered deliberation directed toward treaty lawmaking, the conditions in which these two excusing conditions might apply seem highly unlikely.

ILLEGITIMACY

Illegitimate norms have no moral authority to command our allegiance. So, the presumption that previously agreed upon norms for deliberation morally should be followed would not exist if the norms were morally illegitimate. An account of the legitimacy in international legal norms is a major philosophical

---

[34] This distinction between an emergency violation and violation as a reform strategy is discussed in Allen Buchanan, "From Nuremburg to Kosovo: The Morality of Illegal International Legal Reform" in Allen Buchanan *Human Rights, Legitimacy, and the Use of Force* (Oxford: Oxford University Press, 2010), pp. 299–300 and in Robert Goodin, "Toward an International Rule of Law: Distinguishing International Law-Breakers from Would-Be Law-Makers," *Journal of Ethics* 9 (2005): 225–246.

project. I suggest that, for present purposes, we can avoid that project by thinking in the negative about familiar forms of illegitimacy.

International norms may be illegitimate for at least six different reasons. First, the states that are the principal parties to the norms may suffer from legitimacy problems. Second, the processes by which the norms gained allegiance may be insufficiently voluntary. Third, the continued observance of the norms may be owing to the fact that weaker states have no recourse but to accept the demands of the stronger. Fourth, there may be sufficiently severe negative externalities produced by the observance of the norms. For example, the observance of the norms may profoundly harm the important interests of persons who are not citizens of states that are parties to the norms.[35] Fifth, there may be debilitating constraints on their implementation. If there are no effective institutional means for implementing a norm, or if the moral costs of implementing it are too high, the norm could not be morally authoritative. Finally, norms may lose their legitimacy as the context changes and the nature of global problems changes. For example, a norm of extensive state sovereignty might need to be altered because of global climate change.[36]

Is the right to sustainable development in the UNFCCC deliberative context plausibly illegitimate if we assume these six problems as paradigmatic of the illegitimacy of international norms? First, there are 194 parties to the UNFCCC. A great many of these are states that fail to satisfy plausible liberal democratic criteria for legitimacy, including democratic contestation of political offices, protections of civil rights, and gender equality. But all of the major liberal democracies, including new ones such as South Africa, have ratified the Convention. It is not, then, a treaty that owing to its content appeals particularly to authoritarian and antiegalitarian states. Second, the conditions under which weaker states agreed to the Convention allowed for acceptance or rejection without significant penalty. Unlike the motivation for World Trade Organization membership, there are various prospects for human development outside of the treaty. Third, the practice of deliberation leading up to and during each of the COPs does not seem to exhibit the dynamic of weaker states simply acceding to the demands of the powerful ones. A variety of strategic alliances have come to be and there are significant disagreements between the stronger states, such as between the United States and the European Union.

---

[35] For a discussion of three of the first four possibilities, see Allen Buchan and Robert Keohane, "The Legitimacy of Global Governance Institutions," in Allen Buchanan *Human Rights, Legitimacy, and the Use of Force* (Oxford: Oxford University Press, 2010), pp. 111–114.

[36] Shue, "Eroding Sovereignty."

Fourth, the creation of significant negative externalities is not necessary for respecting sustainable development. The people most negatively affected by development-induced climate change are members of future generations, but it is precisely their interests that sustainable development is meant to protect. Fifth, there do not appear to be institutional barriers to implementing the right to sustainable development. Although recognizing the right to sustainable development will require economic costs for the highly developed states, these do not appear to require bearing high moral costs because the states of the developed world can absorb the economic costs of clean energy production without human development setbacks.[37] And finally, the license to pursue human development is part of a package of powers of sovereignty. Although insistence on the enjoyment of the powers of sovereignty in all traditional state domains is problematic for both environmental and global justice reasons, the deliberations within the constraints of the Convention is part of a well-entrenched process in which sovereign states seek concessions from each other regarding enjoyment of their powers of sovereignty. This process is not likely to die away any time soon. Therefore, considering the Convention in light of these six kinds of illegitimacy suggests that it would be implausible to claim that the right to sustainable development is somehow illegitimate.

DEEP AND SUPERFICIAL JUSTIFICATIONS

I assume that moral justification should be based on considerations reasonable for others to accept.[38] A case for justice in assigning mitigation and adaptation responsibilities can be difficult to make when people hold views about climate change based on misunderstandings of the science of climate change, or are the result of putting too much trust in the accounts of climate skeptics.[39] One response to such intransigence is to review the consensus among climate scientists. Another is to debunk the skepticism by revealing its corporate

---

[37] For an estimate of the costs to the United States of limiting CO_2 emissions to 83 percent below 2005 levels by 2050 – a decent start, but not enough – see *CBO Economic and Budget Issue Brief*, November 23, 2009, p. 12. http://www.cbo.gov/ftpdocs/104xx/doc10458/ 11-23-GreenhouseGasEmissions_Brief.pdf (accessed November 1, 2012).

[38] This is a view of justification derived from John Rawls, *A Theory of Justice* rev. ed. (Cambridge, MA: Harvard University Press, 1999), pp. 506–514.

[39] On the extent of the consensus among climate scientists, see Peter T. Doran and Maggie Kendall Zimmerman, "Examining the Scientific Consensus on Climate Change," *EOS, Transactions, American Geophysical Union* 90 (2009): 22–23. http://tigger.uic.edu/~pdoran/ 012009_Doran_final.pdf (accessed November 1, 2012).

sponsorship.[40] These efforts are important for distinguishing disinterested science from interested public relation efforts.

Another hurdle arises if a justification is based on a comprehensive account of justice that applies to all of the most important aspects of global justice. I call such justifications *deep justifications*. There is currently little philosophical consensus about matters of global justice. Although some accounts of global justice probably contain genuine moral insights and are important parts of our moral understanding of the world, it is doubtful that they are capable of generating sufficient acceptance to guide international policy given current philosophical controversies. Moreover, there is considerable urgency to secure an international climate change agreement. The costs of achieving mitigation, for example, increase as we prolong initiating emissions reductions because in the meantime, the stocks of greenhouse gasses in the atmosphere continue to increase. The longer we wait, the more costly it becomes to stabilize atmospheric concentrations of greenhouse gasses at acceptable levels.

These problems do not beset the kind of argument that I have been making in this chapter. This argument is superficial in the sense that it does not appeal to possibly insightful, but complex and controversial accounts of global justice. On the contrary, it appeals to a familiar sense of the reasonableness of acting in good faith. It is reasonable to keep one's commitments – barring especially strong moral reasons not to do so.

In a previous section I discussed reasons to believe that improved health and education outcomes and greater income are important goods. One might try to defend the right to sustainable development simply and directly by claiming that these goods provide compelling moral reasons to respect the right to human development. But this would require an additional argument. I will have more to say about such an argument shortly. For now, however, recall the normative gap from Chapter 2: The mere claim that something is good does not establish that others have an obligation to promote it. It might be good if I were to help my neighbor with his yard work, but that alone does not entail that I have a duty to help. If, however, I have promised to do so, then, barring a legitimate excuse, I have such a duty. The duty arises independently of the goodness of the helping action. Ordinarily, I have a moral reason to keep my promises that is independent of the content of the promise. Promises obligate promise makers. The prior agreement of Parties to the Convention expressed in the recognition of the right to sustainable development functions the same way. Put simply, it obligates. That the Parties to the Convention have explicitly

---

[40] For an exposé of the misinformation campaign of some climate skeptics, see Naomi Oreskes and Erik M. Conway, *Merchants of Doubt* (New York: Bloomsbury, 2010), chp. 6.

agreed to respect the right to sustainable development is a compelling reason for doing so. That what they have agreed to do promotes a great human good is also morally significant for reasons that I turn to in the next two sections, but a duty of the Parties to respect the right to human development can be justified by the prior agreement alone, just as long as what they have agreed to is not morally odious.

UNREASONABLENESS AND IMPORTANCE

One might object that the charge that an agent is being unreasonable is not a sufficiently strong moral condemnation to capture what is wrong with flouting the right to sustainable development. Perhaps it is not all that important to avoid acting unreasonably. To respond to this objection I contend that the gravity of the charge of being unreasonable in a deliberative context is dependent on the purpose of the deliberation and the party's role in achieving that purpose. We might be exasperated with someone who confounds our plans to have supper as a group at a restaurant by constantly changing his availability, but we should not think the person culpable of grave immorality. After all, we may have an enjoyable – perhaps even a more enjoyable – meal without him. Neither the larger purpose nor the person's role in it is world historical.

In contrast to the supper example, the Convention's purpose to "prevent dangerous anthropogenic interference with the climate system" is exceedingly important. The consequences of unmitigated climate change are likely very severe. And the purpose of allowing human development is equally important. Additionally, the role of some parties in the effort to mitigate climate change is especially important. The Convention prominently recognizes an important role for developed countries. "[T]he developed country Parties should take the lead in combating climate change and the adverse effects thereof."[41] Leading in the negotiations to combat climate change is an expression of taking the purpose of the Convention seriously. Moreover, as developed countries, they have greater capacity to absorb the costs of mitigating and planning for adaptation without serious human development losses. But most importantly, an effective regime of lowering global emissions depends crucially on the participation of the states with very high total emissions; many of these are developed states. The United States is the second-largest emitter, Japan fifth, Germany sixth, Canada seventh, and the United Kingdom eighth. But the role of several developing countries is important in climate change mitigation

41 UNFCC, Art. 3, para. 1.

as well. China is the largest total emitter and India is fourth.[42] Given the great importance of both climate change mitigation and human development, the charge that one of these important parties was unreasonably undermining the process would be an especially grave moral condemnation.

FAIRNESS

The duty to respect the right to human development, according to the superficial justification, is a kind of promissory obligation. This argument might be criticized by someone on the grounds that it is altogether too conventional. What if the parties had never agreed to insert the relevant clause in the Convention? Or suppose there were no Convention; would there be no duty to respect the right to sustainable development? The criticism confuses a sufficient condition of a justified duty – namely, the prior agreement of the parties, with a necessary one. The superficial justification does not rely on the idea that there are no other justifications of the right, even though I have cast doubt on the utility of deep justification on grounds of global justice given the task at hand. Another powerful reason to support the right is based on the value of fairness.

The pursuit of climate change mitigation and adaptation is tremendously morally important in light of the credible predictions of terrible suffering and devastation if mitigation and adaptation are insufficient. In order to achieve this goal, states must engage in international cooperation. But the very concerns that establish the need for cooperation place constraints on it, if it is to be fair. It would be unfair to establish terms of cooperation that will also perpetuate tremendous suffering by means of prolonging underdevelopment. Absent access to inexpensive energy, human development will be slowed. So, terms of cooperation in combating climate change that raise the costs of obtaining energy for least-developed and many developing states would be unfair. Like the defense of the right to sustainable development on grounds of the reasonableness of keeping prior agreements, the present argument seeks to draw the moral norm out of the effort to cooperate internationally to combat climate change rather than applying moral principles that are completely external to the effort. The hope is that such an approach makes sense to participants engaged in the cooperative effort. But there is no need to deny the possibility of insightful external moral approaches.

---

[42] See Union of Concerned Scientists, "Each Country's Share of CO_2 Emissions." http://www .ucsusa.org/global_warming/science_and_impacts/science/each-countrys-share-of-co2.html (accessed November 1, 2012).

Historically, there have been philosophers who have maintained that fairness to states alone matters, just as fairness to persons does, because states are bearers of moral entitlements that are entirely independent of persons. A paradigmatic expression of this view in the modern period is by Emerich de Vattel: "We must therefore apply to nations the rules of the law of nature, in order to discover what their obligations are, and what their rights: consequently the *law of nations* is originally no other than the *law of nature applied* to nations."[43] But such statist views seem widely counterintuitive to anyone who takes seriously that human rights are claims that persons can have against states.

A more promising basis for justifying the fair access to energy among states seeking to mitigate and adapt to climate change is that such fairness matters because of morally important concerns to individuals. In justifying the antipoverty principle in Chapter 1, I argued that a principle that unnecessarily prolonged poverty would be unacceptable to anyone seeking agreement about principles based on the respect for human dignity. A principle that requires policies that hinder the pursuit of energy-intensive macroeconomic policies in least-developed and many developing states would, if followed, prolong poverty for billions of present and future people. If there are alternatives to such policies consistent with climate change mitigation, then mitigation plans based on a principle that requires poverty-prolonging policies are deeply unfair. The antipoverty principle is the basis for claiming that principles, and the policies justified by them, which are inconsistent with the right to sustainable development are deeply unfair to poor people living in poor states.

THE RIGHT TO SUSTAINABLE DEVELOPMENT AND HUMAN RIGHTS

The arguments of the previous sections support two conclusions. First, in the absence of technological breakthroughs lowering the price of energy generated from renewable sources, the right to sustainable development requires that all least-developed and many developing states either be able to increase their CO_2 emissions or have their purchases of renewable energy subsidized by the states with high human development. Second, within the context of the UNFCCC deliberations, proposals that do not respect the right to sustainable

---

[43] Emerich de Vattel's, *The Law of Nations, Or, Principles of the Law of Nature, Applied to the Conduct and Affairs of Nations and Sovereigns, with Three Early Essays on the Origin and Nature of Natural Law and on Luxury* (1797), Preliminaries §6. The Online Library of Liberty ed. p. 31. http://files.libertyfund.org/files/2246/Vattel_1519_EBk_v6.o.pdf (accessed June 25, 2013).

development are unreasonable and deeply morally problematic given the importance of reaching an agreement.

The right to sustainable development is a group right because the pursuit of development is a collective act. Development is something that states – not individuals – pursue. Some philosophers, jurists, and policy makers find group rights to be suspect.[44] Sometimes this is caused by a concern about rights inflation, which could devalue the currency of other rights; other times it is owing to their character as group rights, which might be suspicious on liberal individualist grounds. An adequate defense of the right to sustainable development must respond satisfactorily to both of these concerns.

First, the defense of the right to sustainable development on grounds of the reasonableness of keeping morally significant promises or the importance of fair access to energy is not necessarily a defense of the right as a human right.[45] The right protects the liberty of states to pursue development, and that liberty is tremendously important in the climate change context. But the defense of the right simply takes no position on the matter of whether it is properly a human right. That it would show reprehensible disrespect for the other parties in the treaty negotiating process to violate a fundamental norm that guides the process does not establish that there is a human right that is violated when such disrespect is expressed. When we break promises, we typically do something wrong, but the wrong is not usually thought of as a human rights violation. It might be different in the present case because of the moral importance of the project of human development, but the argument does not depend on it being different. When states act on unfair terms of cooperation and thereby prolong poverty, it seems plausible that prolonging poverty is a human rights violation, as I explored in Chapter 1. But the antipoverty principle does not depend on it being such a violation. So, if there is really a danger to the value of important human rights or necessarily a threat to individual rights by postulating a group human right, then the right to sustainable developed as justified in these two ways should not be guilty of these charges.

Second, the right to sustainable development is consistent with the promotion of individual human rights. Human development widely expands the opportunities for persons in modern societies in which individuals require access to education and health services and opportunities for reliable income. There are, of course, certain religious orientations that encourage voluntary

---

[44] See Jeremy Waldron's "Can Communal Goods be Human Rights?" in his *Liberal Rights: Collected Papers 1981–1991* (Cambridge: Cambridge University Press, 1993), pp. 339–369.

[45] At the very least, group rights are not human rights "in the standard sense" as Nickel puts it. See *Making Sense of Human Rights*, p. 164.

poverty, but this is not the same as advocating involuntary limited access to income.

Electrification is centrally important to human development. According to the UNDP, the indoor pollution caused by the burning of wood and animal dung for cooking results in 1.5 million deaths per year, mostly children under the age of five. This is more than the annual global deaths caused by malaria and is nearly equal to those caused by tuberculosis. "[I]n Bangladesh, rural electrification is estimated to increase income by 11 percent – and to avert 25 child deaths for every 1000 households connected."[46] The burning of fossils fuels is, for many countries, the most economical form of electricity production given current technological capacity and market pricing. Electrification, generated by means of burning fossil fuels, propels human development.

The right to sustainable development protects important interests in education, health, and income. In so doing, the right is consistent with individual human rights that do the same. Article 25, paragraph 1, of the Universal Declaration of Human Rights states that

> [e]veryone has the right to a standard of living adequate for the health and well-being of himself and of his family, including food, clothing, housing and medical care and necessary social services, and the right to security in the event of unemployment, sickness, disability, widowhood, old age or other lack of livelihood in circumstances beyond his control.[47]

Article 26, paragraph 1, holds that "[e]veryone has the right to education. Education shall be free, at least in the elementary and fundamental stages."[48] And Article 26, paragraph 2, affirms that "[e]ducation shall be directed to the full development of the human personality and to the strengthening of respect for human rights and fundamental freedoms."[49] The assignment of a group right of sustainable development to states does not appear to undermine or cheapen the value of important individual rights. On the contrary, it can serve the promotion of these individual rights.[50]

There seems little reason to believe that recognizing the right to sustainable development will be in conflict with important individual human rights. But to reiterate the point made at the beginning of this section, even if the right to

---

[46] UNDP 2007–2008, p. 45.

[47] Universal Declaration of Human Rights, Art. 25, para. 1. http://www.un.org/en/documents/udhr/index.shtml (accessed October 5, 2012).

[48] *Ibid.*, Art. 26, para. 1.

[49] *Ibid.*, Art. 26, para. 2.

[50] See also Waldron, "Can Communal Goods be Human Rights?" pp. 362–364.

sustainable development is broadly consistent with individual human rights, it does not follow that it is a human right in the standard sense.

THE RIGHT TO SUSTAINABLE DEVELOPMENT AND INTERNATIONAL PARETIANISM

Now, if we suppose – although the justifications offered do not require us to do so – that the right to sustainable development is not a human right, we will have to respond to the following challenge: Perhaps the right is not strong enough to sustain the heavy criticism from those who seek to discard it in pursuit of efficiency in climate change mitigation.[51] Eric A. Posner and David Weisbach dismiss the right to sustainable development in the course of defending their conception of efficiency that they call "international Paretianism."[52] If the right to sustainable development is not a fundamental human right, then is there sufficient moral reason to respect it when other important values, such as the efficient use of resources, are at stake? I believe that a compelling case for an affirmative answer to this question can be made. But before presenting it, a few clarifications of Posner and Weisbach's conception of efficiency are necessary.

First, one might think that an efficient climate change–mitigation policy simply is a policy in which everyone is better off in comparison to some set of alternative cases. But Posner and Weisbach limit their analysis to the level of states. Therefore, international efficiency would seem to be a matter of whether all states are better off in comparison to some set of cases. But then another matter arises. This way of characterizing efficiency is sensitive to the facts of resource use; namely, whether it makes states better off, but insensitive to people's fallible beliefs about those facts. As it turns out, this is not Posner and Weisbach's conception of international Paretianism. Their conception of efficiency is sensitive to people's beliefs, not to the facts. To satisfy international Paretianism, "all states must believe themselves better off by their lights as a result of a climate treaty."[53] As long as being better off is measured by some objective consideration, such as national income or gross domestic product, states could wrongly believe that a policy would make them better off, when in fact it does not. In that case, the policy would satisfy international Paretianism,

---

[51] Thanks to Henry Shue for getting me to see the force of this objection.

[52] See Eric A. Posner and David Weisbach, *Climate Change Justice* (Princeton, NJ: Princeton University Press, 2010), pp. 73–98. They criticize climate change proposals based on the right to sustainable development as involving "redistribution," which would be better treated as a separate matter.

[53] *Ibid.*, p. 6.

but in fact would be inefficient. Finally, the standard that all the states believe themselves to be better off is ambiguous unless the comparison set is specified. Must the states believe themselves to be better off than under business-as-usual? Or must they believe themselves to be better off in comparison to all other proposals? If Posner and Weisbach mean the former formulation, their standard is one of believed Pareto Superiority to business-as-usual, and there could be many different policies distributing the burdens of mitigation in various ways that would satisfy it. If the latter, then the standard is one of Pareto Optimality, and only one policy could satisfy it.[54] The formulation quoted earlier has the states believing themselves to be better off "as a result of a climate treaty." This suggests that international Paretianism is a conception of Pareto Superiority to business-as-usual because the indefinite article in "a climate change treaty" would allow for there to be other treaties. Moreover, at another point when Posner and Weisbach are attaching hypothetical costs to mitigation, the comparison they employ is to costs under business-as-usual.[55] It would seem, then, that international Paretianism is a standard of Pareto Superiority for states in comparison to a business-as-usual scenario.

Now, to begin the response that there is sufficient reason to safeguard sustainable development even at the cost of international Paretianism, let's assume we are comparing two policies – both of which would achieve sufficient climate change mitigation, but only one of which would satisfy international Paretianism, and only the other of which would respect the right to human development. This assumption serves to focus the discussion on just these two considerations. In reality, it is possible that satisfying the right to sustainable development could also be Pareto Superior to business-as-usual. More than that, it actually seems probable. With the expectation of unmitigated climate change producing an equilibrium global mean temperature in excess of 4°C, the very high costs are likely to accrue even to highly developed states. Hence, a climate treaty that reduces those costs and safeguards human development should be possible. So, the purpose of this assumption is merely to isolate the comparison between international Paretianism and the right to sustainable development to appreciate the better reasons for safeguarding the latter.

Both the reasonableness of keeping prior commitments and the fair terms of cooperation between states are of moral value. In contrast, efficiency has no independent moral value. Weisbach and Posner concede that international

---

[54] State of affairs A is Pareto Superior to B if at least one party in A is better off than in B and no party is worse off. State of affairs A is Pareto Optimal if there is no alternative in which at least one person is better off and no one is worse off.

[55] See the comparisons of countries R and P at *Climate Change Justice*, pp. 93–94.

Paretianism "is not an ethical principle but a pragmatic constraint."[56] Efficiency is valuable only insofar as the ends we pursue are valuable; it is merely instrumentally valuable. It has no moral value if our end is not morally valuable. Using fewer resources to murder a person is not better than using more. So, although efficiency can provide us with reasons to direct our actions in various ways so as not to waste resources in the pursuit of our values, it cannot provide us with moral reasons against the pursuit of what is morally valuable. Parties to the Convention have a moral duty to respect the right to sustainable development on grounds of the prior commitment to do so. And in any case, there is a moral duty to establish fair terms of cooperation that would bind states even in the absence of the Convention. Violating these duties is proscribed by morality. To fail to efficiently pursue a morally valuable end is not. Morality provides us with reasons to honor the right to sustainable development, even if doing so would not satisfy international Paretianism, but there are no reasons independent of what we value to pursue efficiency.

The previous argument is even stronger than it might appear. Because international Paretianism is belief sensitive and not fact sensitive, it is possible that the policy consistent with the right to sustainable development also, in fact, would be Pareto Superior to business-as-usual, but that international Paretianism would not (because the beliefs of states about efficiency do not necessarily accurately track the facts about efficiency). In that case, the comparison would be between one policy sufficient to achieve mitigation that is both efficient and required by the duty to respect the right to sustainable development, and another policy that is neither efficient nor respectful of the right to sustainable development, but which is believed to be efficient. The right decision in that case would be a no-brainer.

Recognition that Posner and Weisbach's conception of efficiency allows for the possibility of a no-brainer choice raises the question of why they employ a belief-sensitive conception of efficiency rather than a fact-sensitive one. The reason, I conjecture, is that most of the weight of their argument against the right to sustainable development is supposed to be carried by feasibility considerations rather than moral ones. They consider the right to sustainable development to require international redistribution.[57] And they take international redistribution to be politically infeasible, as demonstrated by the failure to achieve a multilateral foreign aid treaty.[58] Moreover, they claim that "history

---

[56] *Ibid.*, p. 6.
[57] *Ibid.*, p. 73.
[58] *Ibid.*, pp. 86–87.

supplies very few cases where states act against their own perceived interests in order to satisfy the moral claims of other states."[59] Their reason for employing a belief-sensitive conception of efficiency is that what matters for the motivation of states are not the facts, but what states believe to be the facts.

One straightforward way to characterize Posner and Weisbach's argument against the right to sustainable development, then, is that international Paretianism is superior on feasibility grounds because it selects solutions perceived to be in the interests of all states, but the right to sustainable development, which licenses making heavier demands on highly developed states, selects solutions not in the interests of highly developed states. Here the consideration is not with actual efficiency, but with what can be agreed upon. However, there is no reason to believe that international Paretianism has advantages on grounds of feasibility since least-developed and developing states will never agree to an international treaty that requires them to sacrifice their legitimate development aspirations.[60] So, even on international Paretianism's home field of feasibility rather than morality, it does not possess any advantages. A more detailed (but brief) philosophical analysis of the comparison between the right to sustainable development and international Paretianism can be found in Appendix C.

LIMITATIONS

The right to sustainable development constrains mitigation and adaptation policy making generally. This argument in defense of the right neither stipulates the exact burdens on any particular state nor specifies precisely the institutional requirements for satisfying the right. Presumably there are multiple arrangements for the distribution of the burdens of mitigation and adaptation that would satisfy it.[61] I am not, then, arguing for a detailed blueprint. In the

---

[59] *Ibid.*, p. 6.
[60] See also J. Timmons Roberts and Bradley C. Parks, *A Climate of Injustice: Global Inequality, North-South Politics, and Climate Policy* (Cambridge, MA: The MIT Press, 2007) chp. 5, but esp. p. 152.
[61] Various widely different candidates can be found in the following writings: See Paul Baer, Tom Athanasiou, Sivan Kartha, and Eric Kemp-Benedict, *The Greenhouse Development Rights Framework: The Right to Development in a Climate Constrained World* (Berlin: Heinrich Böll Foundation, Christian Aid, EcoEquity, and the Stockholm Environment Institute, 2008), rev. 2nd ed. http://gdrights.org/wp-content/uploads/2009/01/thegdrsframework.pdf (accessed November 9, 2012); Tariq Banuri and Niclas Hällström, "A Global Programme to Tackle Energy Access and Climate Change," *Development Dialogue* (2012): pp. 264–279; and Humberto Llavador, John E. Roemer, and Joaquim Silvestre, "North-South Convergence and

context of complex negotiations, it is a virtue of the norm that it provides moral guidance without very specific requirements. This leaves room for considerable negotiation about the means of respecting the right.

We face the threat of an explosion in carbon emissions without a concerted international effort to reduce emissions. Unless the demand for fossil fuels is diminished by means of regulation that effectively puts a price on carbon emissions, the global consumption of coal is expected to rise approximately 50 percent by 2035, with almost all of this growth occurring in Asia. During this time period, China's use of coal is expected to increase 55 percent and its number of coal-fired plants is expected to double.[62] The right to human development should not be construed as an unlimited entitlement to emit CO_2; that would be an ecological disaster.[63] And as noted at the end of Chapter 4, the morally constrained CO_2 emissions budget allows us just another half-trillion tons of carbon emissions. CO_2 emissions will have to be drastically reduced very soon, and eventually halted completely. So, a right to emit CO_2 indefinitely is simply inconsistent with mitigating climate change. Clearly, emissions reductions need to begin immediately in the highly human-developed states, but eventually the least-developed and developing states will need to have carbon-free economies as well.

Within a treaty context designed to avert dangerous climate change, the right to sustainable development is a liberty for least-developed and developing states to pursue human development without additional energy costs caused by climate change mitigation policies. The role that this right plays in the international context is analogous to the role that individual rights play in the domestic. It protects the liberty of the rights holders against the interfering actions of other parties and against overreaching by the common political authority. The protection of this liberty of states might require access to coal to burn for electricity for a limited time, and it will probably also require subsidies for the generation of electricity by means of renewable sources.[64]

the Allocation of CO_2 Emissions," unpublished, http://www.tiger-forum.com/Media/pdfs/SUSTAIN/John%20Roemer.pdf (accessed June 26, 2013).

[62] U.S. Energy Information Administration, "International Energy Outlook 2010 – Highlights." http://www.eia.doe.gov/oiaf/ieo/highlights.html (accessed November 1, 2012).

[63] Tim Hayward argues against entrenching a right to emit CO_2 in "Human Rights versus Emissions Rights," *Ethics and International Affairs* 21 (2007): 431–450. This point is crucial. I have not, however, adopted his alternative framework of an equitable distribution of ecological space.

[64] See Banuri and Hällström, "A Global Programme to Tackle Energy Access and Climate Change."

In this chapter I have pursued an important purpose that moral philosophy can serve, namely, greater understanding of our collective commitments. Moral philosophy can help us to better understand what we have good reasons to do. As important as it is to arrive at that understanding, it is much more important to act on those reasons.

6

Responsibility and Climate Change Policy

"Seemeth it a small thing unto you to have eaten up the good pasture, but ye must tread down with your feet the residue of your pastures? And to have drunk of the deep waters, but ye must foul the residue with your feet? And as for my flock, they eat that which ye have trodden with your feet; and they drink that which ye have fouled with your feet."

– Ezekiel 34:18–19

There are costs in taking morally appropriate precautions against climate change–caused future catastrophes. These costs involve expenditures in the pursuit of climate change mitigation and savings to reduce future adaptation costs. Many of us need to consume energy generated by means other than burning fossil fuels. For some time at least, this will be more expensive than using coal, oil, and natural gas. If we divide all of our income into two broad categories – one directed toward consumption and another directed toward savings – spending more on electricity generated by, say, wind or solar cells will result in less money available to use for other kinds of consumption. This can be thought of as a kind of savings, but not primarily for ourselves. Most of the beneficiaries of climate change mitigation will live after we have died. Increased energy prices and tax-financed public investments in adaptation amount to savings programs for the future forced on us by current public policy. But for many people now alive – indeed, for most of them – savings for others would simply be a means by which their impoverishment would be deepened or prolonged or both.

The life prospects for nearly half the world's population are bleak and for a majority among them, the circumstances are wretched. Weak educational infrastructure, poor health care, and little income conspire to produce short lives with few alternatives other than almost no regular income or employment under brutally difficult conditions with only meager pecuniary reward.

However, recent experience shows us that improvements in education, longevity, and income are possible. In the second half of the twentieth century, several states climbed out of generalized misery to relatively high levels of human development. There are debates about the best strategies for doing so, and about the background conditions that make this most likely, but there is no doubt that the process is tremendously energy intensive. Inexpensive sources of energy make human development more affordable. And given current technology, the cheapest sources are fossil fuels; coal, in particular.

The right to sustainable development defended in Chapter 5 is a liberty that certain states may claim to pursue energy intensive human development despite the duties of intergenerational justice, which the present generation has to mitigate claim change. This liberty is a matter of justice. Justice and responsibility are conceptually related. They may be looked at as two different sides of the same moral coin. Justice concerns which people are owed what. In the words of John Stuart Mill, it involves "a claim on the part of one or more individuals, like that which the law gives when it confers proprietary or other legal rights."[1] Responsibility concerns which people owe what. Another way of putting it is that in virtue of a claim of justice, persons are moral creditors who may call in what is rightfully theirs. In virtue of being responsible, people are moral debtors, who must give up what is not rightfully theirs. It is no surprise, then, that similar moral concerns, such as fairness, often apply to both justice and responsibility.[2]

In this chapter I focus on two concerns of responsibility as it relates to climate change policy. The first part of this chapter is directed to the question of why our generation, comprising a small subset of all polluters historically,

---

[1] John Stuart Mill, "Utilitarianism," in J. M. Robson ed., *John Stuart Mill: Essays on Ethics, Religion and Society* (Toronto: University of Toronto Press, 1969), p. 247. http://files.libertyfund .org/files/241/0223.10_Bk_Sm.pdf (accessed October 30, 2012).

[2] Explicating justice and responsibility in terms of creditors and debtors goes back at least to Friedrich Nietzsche, who does so for very different purposes in Keith Ansell-Pearson, ed., and Carol Diethe, trans., *On the Genealogy of Morality* (Cambridge: Cambridge University Press, 2006). See, for example: "The feeling of guilt, of personal obligation, to pursue our train of inquiry again, originated, as we saw, in the oldest and most primitive personal relationship there is in the relationship of buyer and seller, creditor and debtor: here person met person for the first time, and *measured himself* person against person" (p. 45). Nietzsche has in mind an historical conception of justice, which I reject for the purposes under discussion in this chapter. Martin Luther King Jr. uses similar language in his "I Have a Dream" speech: "In a sense we have come to our nation's capital to cash a check. When the architects of our republic wrote the magnificent words of the Constitution and the Declaration of Independence, they were signing a promissory note to which every American was to fall heir. This note was a promise that all men – yes, black men as well as white men – would be guaranteed the unalienable rights of life, liberty, and the pursuit of happiness." http://www.ushistory.org/ documents/i-have-a-dream.htm (accessed December 13, 2013).

is morally responsible to mitigate climate change. Why would we be subject to appropriate and trenchant moral reproach, including self-reproach, if we failed to do what seems necessary to avoid dangerous climate change? After discussing that, I turn to the issue of the assignment of mitigation responsibilities among states. On the basis of what moral considerations should a climate treaty distribute the responsibility to assume the costs of climate change mitigation and adaptation? The Convention contains language concerning this question and the language has been the object of much policy discussion. In Article 3, paragraph 1, the Convention proclaims: "The Parties should protect the climate system for the benefit of present and future generations of humankind, on the basis of equity and in accordance with their common but differentiated responsibilities and respective capabilities."[3] I offer an interpretation of this language that, I believe, makes the best moral sense of it.

THE PRAGMATICS OF RESPONSIBILITY

As a moral concept, responsibility picks out those persons who owe a debt, and conceptions of responsibility will vary according to the understanding of the debt.[4] There is much disagreement among philosophers writing about responsibility regarding what triggers it and why. Sometimes this is substantive disagreement (some of which I discuss in this chapter), but not always. Philosophical disagreement about responsibility can often be understood as philosophers talking – literally – at cross purposes. Any conception of responsibility will be compelling to the extent that it fulfills in a satisfactory way the purpose for which it is developed. There are several legitimate purposes that a conception of responsibility can serve. Appreciating this facilitates clarity on the points at which there is real substantive disagreement; namely, about which is the best account of responsibility given the purpose that the conception serves. Pragmatism in philosophy is often associated with the view that the purposes of a conception under discussion are relevant to its assessment. In that sense, this section offers a pragmatic approach to responsibility.

One important purpose of a conception of responsibility is to identify whom we should morally praise or blame for an action or outcome. Conceptions of moral responsibility are evaluated in light of this purpose. The legal

---

3 The United Nations Framework Convention on Climate Change. http://unfccc.int/files/ essential_background/background_publications_htmlpdf/application/pdf/conveng.pdf (accessed November 8, 2012).

4 The distinction between a concept of something and the conception of it derives from John Rawls. See his *A Theory of Justice*, rev. ed. (Cambridge, MA: Harvard University Press, 1999), p. 9.

analog identifies who deserves legal sanction. Not everyone who is legally responsible is morally responsible because legal guilt, as determined by due process of the law, does not necessarily track moral guilt. And the morally guilty can, of course, be found legally innocent, as the critics of due process decry.

Another use of a conception of responsibility is to identify some kind of link between a person and an outcome about which the person owes a moral debt. Something happens and we assert that the person is responsible for its happening. The legal scholar Tony Honoré calls this "outcome responsibility," and describes it as follows: "Outcome responsibility means being responsible for the good and harm we bring about by what we do. By allocating credit for the good outcomes of actions and discredit for bad ones, society imposes outcome responsibility; though often the rewards it attaches are outside the law, and the sanctions it imposes are informal and vague."[5] This is not fully adequate because it ignores outcome responsibility for what we fail to do. Given another run at it, Honoré puts it this way: "[W]e are, if of full capacity and hence in a position to control our behavior, responsible for the outcome of our conduct, whether act or omission."[6] A difficulty for conceptions of outcome responsibility is accounting for the link between the person and the outcome that establishes the credit for the good and discredit for the bad, especially when outcomes include those that come about because we fail to act.

The philosopher David Miller distinguishes outcome responsibility from causal responsibility as follows: "We ask about causal responsibility when we want to know why something happened. In the case of outcome responsibility, our interest is different. We want to know whether a particular agent can be credited or debited with a particular outcome."[7] According to Honoré, the outcomes for which we are responsible in light of a failure to act include only omissions, which he understands to be characterized as more than simply not acting, but also violating a norm by failing to act.[8] Because this is the business of crediting and debiting, outcome responsibility – unlike causal responsibility – is moralized. Sometimes we accrue debt, not for what we have done, but for what we failed to do, when we should have done it. A mechanic who fails to sufficiently tighten the bolts on the engine is outcome responsible for the engine's subsequent problems, despite literally doing nothing.

[5] Tony Honoré, *Responsibility and Fault* (Oxford and Portland: Hart Publishing, 1999), p. 14.
[6] *Ibid.*, p. 76.
[7] David Miller, *National Responsibility and Global Justice* (Oxford: Oxford University Press, 2007), p. 87.
[8] Honoré, p. 77.

Miller contends controversially that "to be morally responsible for something you must be outcome responsible for that thing."[9] This might be too quick. Whether or not we should agree will depend crucially on the details of outcome responsibility. If, for example, outcomes do not include our actions and attitudes, this claim seems implausible. H. L. A. Hart and Honoré rightly note that "[w]e blame a man who cheats or lies or breaks promises, even if no one has suffered in the particular case."[10] There are several kinds of cases of moral responsibility relevant here, including those in which an evil attempt fails; those in which we attempt nothing sinister, but behave negligently and luckily bring no harm on anyone; those in which the character trait displayed by the action, such as dishonesty, is morally repugnant; and those that involve violating an important norm albeit without discernible negative impact, such as driving drunk without causing an accident. Only if we are, in some sense, outcome responsible in cases like these is Miller's claim plausible. Nonetheless, my interest here is not to determine the truth of Miller's claim, but merely to provide some clarity about the different conceptions of responsibility.

The purposes to which we put a conception of responsibility can also include assigning the costs of remedying a circumstance in need of repair. Miller calls this "remedial responsibility." Miller develops a "connection theory" conception of remedial responsibility, according to which a person must be linked to another person's condition in one of several ways in order for the former person to be remedially responsible for repairing the circumstances of the latter.[11] Outcomes may be sufficient to be remedially responsible, according to Miller, but they are not necessary. He does not seem to believe that there is a single necessary condition for the assignment of remedial responsibility. Sometimes a reason that an assignment should fall on a person to repair a situation is just that it is his turn to clean up the mess or that he has more time than everyone else to do it. Miller's discussion of remedial responsibility focuses on discrete actions (A punching B) or particular situations (a search for a lost child), usually occurring within the background of stable assignments of entitlements to property (plowing land adjacent to an apple tree owned by neighbor).

The concept of repair typically suggests the reestablishment of a moral equilibrium that existed prior to an action, event, or a series of actions and events that set things off-kilter; Miller's discussion seems to apply to repairs

9   Miller, *National Responsibility*, p. 89.
10 H. L. A. Hart and A. M. Honoré, *Causation in the Law* (Oxford: Oxford University Press, 1959), p. 59.
11 Miller, *National Responsibility*, pp. 101–104.

of this kind. But repairs can also be needed in circumstances of a different kind, which call for the establishment of a new and fair system of background rules for social dealings. Here the work is less the handyman sort of fixing and more the architectural. The design of a system of responsibility to support a new and better institutional order is needed. This might be called for because a long history of injustices has fundamentally disrupted social relations and made a return to a status quo ante impossible. Apartheid in South Africa was in need of this kind of repair. Establishing fair background rules is also called for when either old forms of interaction accumulate to produce a radically changed circumstance or new forms of interaction emerge. Depending on how novel one thinks the phenomenon of globalization is, international trade can be looked at in either of these ways. In either case, new background rules to govern interaction fairly are called for in light of recent developments. I refer to the architecture of responsibility under the design of new rules as *social responsibility* to distinguish it from Miller's handyman conception.

Social responsibility is distinct from remedial responsibility in two ways. First, social responsibility is not directed to discrete rules that assign responsibility to make repairs when a breakdown occurs in the system of rules and institutions. Rather, social responsibility is directed to the design of a set of institutions. Second, social responsibility is not directed to the activities that would remedy existing social injustice; that is yet another important purpose of responsibility, discussed, for example, by Iris Marion Young.[12] Social responsibility assigns burdens under the blueprint for a new and better system of institutional rules.

GENERATIONAL MORAL RESPONSIBILITY AND REASONS FOR MITIGATION

In Chapter 1 I observed that future persons would have good reasons to avoid the risks of climate perturbations. The risks and uncertainties of climate change threaten catastrophes to people in highly developed as well as least-developed countries. But poor persons in poor countries are especially vulnerable because they have little to no means of adapting to climate change or of insuring themselves against the costs it will exact, and their states have little money to invest in adaptation policies. These are reasons for mitigation. I have also argued

---

[12] See Iris Marion Young's social connection model of responsibility: "Being responsible in relation to structural injustice means that one has an obligation to join with others who share that responsibility in order to transform the structural process to make their outcomes less unjust." See her *Responsibility for Justice* (Oxford: Oxford University Press, 2001), p. 96.

that it would be a serious wrong to impose poverty-prolonging climate change on people now and in the future if there is a mitigation plan at comparatively minor costs. We accept that involuntary poverty is something that everyone has good reason to avoid and it is inconsistent with respect for human dignity to impose a policy based on a principle that would be unacceptable to people seeking agreement about principles on the basis of respect for human dignity. It seems, however, to under-describe the matter to speak of climate change only as a threat that will prolong poverty. Cascading uncertainties about possible catastrophes must figure prominently in our accounting of the wrong of climate change.

Who is morally responsible for stopping the wrongful harm that will be imposed on a great many people if climate change is not sufficiently mitigated? Perhaps the answer seems obvious. It must be us – our generation – taken as a group. We can change our greenhouse gas emissions policies and thereby stabilize the concentrations of greenhouse gases in the atmosphere that have been accumulating for 150 years. There is no one else to do the job.

It can be fruitful to consider objections even to what seems obvious. So, consider this one: Slightly over half the people now alive in the world are younger than thirty years old and just more than thirty years ago, the concentrations of the CO_2 in the atmosphere were already about 85 percent of what they now are. Most people alive now have not contributed a great deal to the total stock of emissions in the atmosphere. If the threshold of atmospheric concentrations beyond which it is virtually impossible to limit warming to around 2°C is breeched, the emissions of the last thirty years will not be a large percentage of the total stock. Suppose we take for granted Miller's assertion about the relationship of outcome responsibility to moral responsibility: We produced only about 15 percent of the greenhouse gases in the atmosphere. Most of us did nothing to produce the other 85 (or more) percent that were already in the atmosphere at the times of our births. Given the share of current atmospheric concentrations that we have produced, our outcome – and, by implication, moral responsibility for climate catastrophes – is minor.

According to the framing of the issue in the previous argument, because our causal role in future suffering is relatively limited; the central moral duty we have – namely, whether to help those in need (or, more accurately, to prevent people from falling into need) – is relatively weak. There are two problems here, one minor, another major. The minor problem is that even if the argument frames the problem in the correct way, it downplays the significance of the moral issue. The fact that we are not the primary cause of a person's suffering does not relieve us of a moral duty to help. And even if we would not be the primary cause of future climate change-related suffering, we would still have a moral duty to take serious steps to prevent it given its gravity. We can allow that our duty to act is limited to actions that do not overburden us, but we are nowhere near exceeding that limit because we are presently doing exactly nothing. Generally, one can be let off the hook of a duty to help by the helping actions of others, but there is no other generation that can prevent the suffering of persons in the future. So, even if the moral question is whether or not to help people in the future, it would seem that we have a duty to take serious steps to prevent future suffering and devastation by mitigating. To fail to do so is to be blameworthy for a morally important omission.

The more serious problem with the argument is that its manner of framing the moral problem ignores a deeply important moral issue. I introduced the idea of a morally constrained CO_2 emissions budget at the end of Chapter 4. Our failure to mitigate climate change would constitute using more than our fair share of a morally constrained intergenerational CO_2 emissions budget; that would not be a failure to help others in need. On the contrary, it would harm people in several ways. First, it would involve taking more of the CO_2 emissions budget than rightly belongs to us. Second, it would undermine the capacity for moral choice of future persons (roughly the next generation) who would be left with the choice of either emitting CO_2 in excess of the cumulative budget – and in doing so, heaping massive risk and uncertainty on subsequent generations – or making a rapid and human-development slowing transition to renewable energy.[13] Third, regardless of which choice the next generation makes, we would be implicated in the subsequent suffering because we framed the choice of that generation.

Our generation, taken collectively, would be morally blameworthy for the severe moral failure of failing to mitigate. That the idea of blaming us makes sense can also be appreciated by considering that had things been different, we might have been excused. Had we made serious efforts to avoid dangerous climate change, by adopting the kind mitigation plan we currently think necessary to avoid dangerous climate change, and had it turned out that climate sensitivity was significantly higher than we had any good reason to believe, our efforts might have failed. But in that case, our failure would seem excusable. Excusing us, of course, would make sense only if, in failing to attempt to mitigate, we became blameworthy.

[13] For a discussion of placing future generations in worse moral positions, see also Henry Shue, "Responsibility to Future Generations and the Technological Transition," in Walter Sinnot-Armstrong and Richard B. Howarth eds., *Perspectives on Climate Change: Science, Economics, Politics, Ethics* (Amsterdam and San Diego: Elsevier, 2005), pp. 272–275.

The argument clarifying the nature of our moral failing (should we not adequately mitigate climate change) appeals to ordinary moral concerns of harm prevention, taking only a fair share, and respecting persons. But the philosopher Dale Jamieson argues that our familiar moral concepts fail in the case of climate change because of peculiar features of the problem.[14] Most of the suffering and devastation that climate change is likely to cause will be in the fairly distant future. Moreover, there will be multiple threads of causation that will result in any particular person suffering. And the persons in the present who are pulling the threads that cause the climate-related harm number in the billions. If a claim of moral responsibility requires establishing the responsibility of particular individuals living now for the suffering of particular individuals in the future, with the parties connected by means of a casual nexus as in the law of torts or delict, its plausibility would be very weak. But the plausibility of the aforementioned arguments is evidence that we do not need to establish links between particular individuals for specific instances of suffering to attribute responsibility to our generation for using more than its fair share of the emissions budget and for undermining the capacity for moral choice of future generations. And, when it comes to attributing generational responsibility for harms, it is enough to be reasonably confident that the incidence and intensity of climate disasters such as droughts, floods, and tropical storms will increase. We need not tie particular living individuals to the suffering of specific future individuals by a discernible causal thread to establish generational responsibility for harm.

Might we legitimately complain that it is unfair that we should have to pay such a high mitigation bill? If we were to complain, who could the complaint be against? None among our generation, since it is we – all of us now living – who are charged with the climate bill. The complaint could only be directed to past generations. It would have to be a complaint that our generation must mitigate because of the emissions of our ancestors, not merely because of our own.[15] They have vanished from the table and left us with a bill to pay. It is, in a certain sense, our bad luck to be left to pay for a mitigation bill

---

[14] See his example of Jack and Jill in Dale Jamieson, "Climate Change, Respect, and Justice," *Science and Engineering Ethics* 16 (2010): 436–437. This is a long-standing concern of Jamieson's. For example, see also his discussion of Jones and Smith in Dale Jamieson, "Ethics, Public Policy, and Global Warming," *Science, Technology, and Human Values* 17 (1992): 148–149. Stephen M. Gardiner also addresses Jamieson's concern in *A Perfect Moral Storm: The Ethical Tragedy of Climate Change* (Oxford: Oxford University Press, 2011), pp. 432–434.

[15] In the course of criticizing a beneficiary-pays principle, Simon Caney presses a criticism like this. The target of the criticism need not be limited to that principle. See his "Environmental Degradation, Reparations, and the Moral Significance of History," *Journal of Social Philosophy* 37 (2006): 476. Perhaps, if the criticism is strong, the view that the present generation must

in excess of our own emissions.[16] But there is no reason to think that such bad luck relieves us of a responsibility. If others fail to stop a wrong, my own inaction is not therefore justified.[17] Nor is it obviously unfair in this case that we should pay. After all, we have not only inherited a bill from our ancestors; many of us are much wealthier than our ancestors were at the beginning of the industrial revolution. We are the recipients of capital that has accumulated over a millennium or more. It is a trend that disturbed some philosophers in days gone by, among them Immanuel Kant.

> What remains disconcerting about all this is firstly, that the earlier generations seem to perform their laborious tasks only for the sake of the late ones, so as to prepare for them a further stage from which they can raise still higher the structure intended by nature; and secondly that only the later generations will in fact have the good fortune to inhabit the building on which a whole series of their forefathers (admittedly, without any conscious intention) had worked without themselves being able to share in the happiness they were preparing.[18]

In light of the many dividends that we enjoy, redounding to us like compound interest from the "laborious tasks" of our predecessors, our bad luck in having to mitigate the emissions of some of them hardly seems like an instance of transhistorical unfairness. On the contrary, we have done well by the bargain, if you can call it that. They may have failed to mitigate a problem that they did not know existed, but they have provided us with the means to tackle the problem at no great costs to ourselves. If we fail to do it, unlike the treasure our ancestors bequeathed to us, we will be passing along easily avoidable misery.

It is the moral responsibility of our generation to mitigate climate change. We have frightfully little time to develop a mitigation plan and to avoid imposing considerable risks and cascading uncertainties on people in the future. If we fail, blame will surely fall to us, just as surely as suffering will befall future people.

---

mitigate, even though the problem is at least partially inherited, is vulnerable to it as well. So, I try to show why, in this case at least, the criticism is not strong.

[16] Henry Shue takes it to be only bad luck and not unfair in an analogous situation in which three people innocently loading a camel are responsible for the fourth person not being able to load it. See Shue, "Historical Responsibility," 3.

[17] This is argued with admirable clarity in Anja Karnein, "Putting Fairness in Its Place: Why There Is a Duty to Take up the Slack," unpublished.

[18] Immanuel Kant, "Idea for a Universal History with a Cosmopolitan Purpose," in Hans Reiss, ed. and H. B. Nisbet, trans., *Kant's Political Writings* (Cambridge: Cambridge University Press, 1970), p. 44.

OUTCOME RESPONSIBILITY EXTENDING INTO THE FUTURE

The effort to mitigate climate change is an intergenerational one. Our morally salutary mitigation policies, if they were ever to come to pass, could be undercut by the disregard of future generations. But in terms of energy policy, there is little reason to worry about that occurring. If our generation effectively made the change to renewable energy, which requires raising the price of fossil fuel consumption, then innovation would drive down the costs of alternative fuels and after a while there would be no significant incentive to go back to them. But population growth and land use could conspire to produce greenhouse gas emissions in the future, even if we were to begin to implement a mitigation plan in the very near future. By 2050 the global population is expected to have grown to 9 billion, about 30 percent more than now. Between 10 and 12 percent of all current greenhouse gas emissions are the result of agriculture practices. Another 17 percent of all greenhouse gases can be credited to deforestation and forest degradation. The hunger of a growing population and the expansion of human settlements could result in greenhouse gas emissions that burden subsequent generations with climate change risks and uncertainties.

In these matters our outcome responsibility seems to extend farther than the emissions of our present generation. Attention to human development now – in particular, through policies that educate and train women and address the financial insecurity of families – should help to arrest population increases after 2050.[19] Those born into impoverished circumstances cannot be outcome responsible for those circumstances; only members of earlier generations are capable of changing the circumstances in which a person is later born. Although the sources of poverty in individual cases will vary considerably from parental neglect to international indifference and many other possible factors, our generation has the power to advance human development for the benefit of future people. We can do this by encouraging good governance, promoting education and health, establishing fair international trade and financial relations, and transferring resources to impoverished people. Such efforts will not prevent parents from acting irresponsibly vis-á-vis their children, but they will improve the background circumstances in which children are born and come to adulthood.

The considerations of the previous paragraph do not entail that members of future generations would bear no outcome and moral responsibility for land-use practices that produced greenhouse gas emissions. To the extent that

[19] This view is defended by, among others, Amartya Sen in *Development as Freedom* (New York: Anchor, 2000).

it is in their power to reduce those emissions, they certainly would bear such responsibility. Still, people born into desperate circumstances of underdevelopment might be morally excused for their emissions, despite being outcome responsible. Given the insufficient attention of our generation to remediate the poverty into which billions of people will be born, we share in the outcome responsibility for future emissions caused by desperate attempts at survival.

ASSIGNING INTRAGENERATIONAL RESPONSIBILITY UNDER A MITIGATION PLAN

The previous arguments establish, I think, that our generation is morally responsible for carrying out an effective mitigation plan that would drastically reduce the likelihood of future climate catastrophes. This includes billions of people in a variety of circumstances. Effective mitigation will require paying more for energy (at least in the short term) because renewable energy is not as cheap as coal and other fossil fuels. And savings to assist in adaptation will divert money that might have been directed elsewhere. In Chapter 5 I argued that the right to sustainable development is a liberty possessed by the least-developed and some developing states not to be assigned costs that would slow the process of human development. How should this generational responsibility be divided intragenerationally within the framework of an international climate change agreement?

A comprehensive international agreement on climate change mitigation and adaptation would establish a new framework for energy use and its costs. One important desideratum of such a framework is effective mitigation and adaptation, but that cannot plausibly be the only one. In recognizing the right to sustainable development, the Convention in effect states another desideratum: A mitigation policy must establish a fair framework for energy use. This framework will lay the burdens of mitigation on certain states and must allow the pursuit of sustainable development in the least-developed and developing states. This section discusses the architecture of social responsibility for such a framework.

Is there a credible account of how to assign responsibilities for climate change mitigation that is consistent with the right to sustainable development? As noted earlier, the Convention holds that "equity" and "common but differentiated responsibilities and respective capabilities" are the basis on which we should determine how much to require of each state. On the face of it, it is unclear how many moral factors are stated here. On one reading there are three: equity, common but differentiated responsibilities, and respective capabilities. But it is also perfectly plausible that equity is meant to be understood in terms of the latter two. Perhaps it seems obvious that the responsibilities are distinct from the capabilities; the section is often read that way. Some take "responsibilities" to refer to a state's share of total historic emissions, and "capabilities" to refer to a state's ability to pay for climate change mitigation and adaptation, in which ability is a function of its gross domestic product or the income of some subset of its members.[20] But it is also possible that responsibilities vary with respective capabilities. This is the reading that makes the best moral sense. I argue that an account of social responsibility for mitigating climate change based on the ability of states to mitigate without delaying human development has two virtues. First, it avoids the various problems of trying to have responsibility track historic contribution for current concentrations of CO_2 in the atmosphere. And second, it coheres nicely with the right to sustainable development.

RESPONSIBILITY TIED TO HISTORIC EMISSIONS

How should the social responsibility to mitigate climate change be assigned? In the context of UNFCCC negotiations, the bearers of this responsibility must be states. The Convention's language of the "common but differentiated responsibilities" of states has led some people to conclude that a morally acceptable international treaty should assign responsibilities to states according to their historic contribution of greenhouse gases, especially CO_2 (because of its long atmospheric residence time), to the atmosphere.[21] This view is similar to a principle from other aspects of environmental policy and ethics known as the polluter-pays principle. The idea is simple: Those who pollute should pay in proportion to their contribution to the overall pollution problem. In many jurisdictions the legal application of the principle is limited to those who culpably pollute. In the case of climate change, the principle is often understood to apply to historical emissions broadly and the contribution is each state's portion of those emissions. In international climate change negotiations, a version of the polluter-pays principle was proposed by Brazil in 1997 in negotiations

---

[20] The Greenhouse Development Rights framework is just one example of reading the language that way. See Paul Baer, Tom Athanasiou, Sivan Kartha, and Eric Kemp-Benedict, *The Greenhouse Development Rights Framework: The Right to Development in a Climate Constrained World* (Berlin: Heinrich Böll Foundation, Christian Aid, EcoEquity, and the Stockholm Environment Institute, 2008), rev. 2nd ed. http://gdrights.org/wp-content/uploads/2009/01/thegdrsframework.pdf (accessed November 9, 2012).

[21] A defense of this view can be found in Stephen M. Gardiner, "Ethics and Climate Change: An Introduction," *WIREs* 54 (2010), 54–66. A complex measure of historic accountability can be found in Roberts and Parks, *A Climate of Injustice: Global Inequality*, pp. 153–173.

leading up to the Kyoto Protocol. The so-called Brazilian Proposal attributes responsibility to mitigate climate change to a country (applicable only to Annex I countries) on the basis of the percentage that they contribute to total global emissions between a stipulated baseline year and a later stipulated year.[22]

The polluter-pays principle might be defended on either fault or no-fault grounds.[23] Fault conceptions of responsibility are at home in the law of torts or delict, which seek to assign responsibility for accidents against a stable background of property entitlements. In order for an agent to be liable in fault for an outcome, typically several conditions must be satisfied, including the following: (1) outcome – the outcome must be credited to the agent; (2) care – the agent's action must have contravened a standard of care, to which it is reasonable to hold an agent; (3) voluntariness – the agent must have acted voluntarily; and (4) knowledge – the agent must have acted with knowledge about the likely outcomes of her action. One kind of defense of a fault conception of responsibility stresses the importance of liberty. In the absence of satisfying these conditions, any sanction unjustifiably invades the agent's liberty. A different kind of defense invokes forward-looking considerations. By assigning responsibility only to those who voluntarily and knowingly create problems, we establish a system of incentives that reduces the incidence of such misdoing and we lower the incidence of externalities – the costs of misdeeds being passed along to others.

Several problems with using fault liability for past emissions are reasonably clear.[24] Consider three such problems. First, in the case of previous generations, negligence or recklessness did not typically occur. Most emissions were not the product of actions that contravene a reasonable standard of care. On the contrary, many people used energy either to produce what they needed to raise themselves and their families out of impoverished conditions or simply to go about the mundane business of getting to work and heating their homes.

---

[22] United Nations Framework Convention on Climate Change, Ad Hoc Group on the Berlin Mandate, Seventh session, Bonn, July 31–August 7, 1997, Implementation of the Berlin Mandate, Addendum, Note by Secretariat, pp. 3–37. http://unfccc.int/resource/docs/1997/agbm/misco1a03.pdf (accessed December 12, 2012).

[23] For more on the application of the distinction between fault and no-fault accounts in climate change policy, see Henry Shue's classic, "Subsistence Emissions and Luxury Emissions," *Law & Policy* 15 (1993): 39–59.

[24] There is a great deal written on this. See, for example, Simon Caney, "Cosmopolitan Justice, Responsibility, and Global Climate Change," *Leiden Journal of International Law* 18 (2005): 747–775 and David Weisbach, "Negligence, Strict Liability, and Responsibility for Climate Change," Discussion Paper 10–39 (Cambridge, MA: Harvard Project on International Climate Agreements, 2010). http://belfercenter.ksg.harvard.edu/files/WeisbachDP39.pdf (accessed November 12, 2012).

Many of these activities we would not want to condemn and some we would want to commend. Second, there is a problem of ignorance. Most of the present stock of CO_2 concentrations was produced by people who were not aware of the damage that their emissions would cause.[25] General knowledge that climate change is produced by greenhouse gases is relatively recent. In 1988 the United Nations General Assembly adopted Resolution 43/53, "[n]oting with concern that the emerging evidence indicates that continued growth in atmospheric concentrations of 'greenhouse' gases could produce global warming with an eventual rise in seal levels, the effects of which could be disastrous for mankind if timely steps are not taken at all levels."[26] It was only that same year that the Intergovernmental Panel on Climate Change (IPCC) was founded by the World Meteorological Association and the United Nations Environmental Programme. The *First Assessment Report* of the IPCC was not issued until 1990. Perhaps in an effort to avoid this problem, the Brazilian Proposal measures emissions only from 1990 onward. Third, a serious practical problem of insufficient collection exists, even if we could establish that people knew what they were doing and breached a reasonable standard of care, because much of the damage-causing stock of atmospheric CO_2 was produced by people who are now long dead.[27] Despite the remarkable advances of modern medicine, there is little prospect of recouping costs from them. So, any assignment of responsibility on the basis of fault necessarily ensures that full costs will not be recouped. This problem is magnified if the living are only assigned fault for their emissions subsequent to 1990. A credible fault account of responsibility would seem to be limited to the emissions of those people currently living and only since the late 1980s or early 1990s. Although this is a means to divide up responsibility within the present generation alone, the problem of insufficient collection leaves responsibility for all earlier emissions unassigned.

Nor are the problems solved if the assignment of responsibility is switched from individuals to states.[28] Assigning fault to states faces at least four severe problems. The first three echo the three problems of assigning fault to individuals. First, there is the problem of a lack of negligence or recklessness in state macroeconomic policy directed toward economic development. Second,

---

[25] For a good discussion of this problem, see Simon Caney, "Climate Change and the Duties of the Advantaged," *Critical Review of International Social and Political Philosophy* 13 (2010): 208–210. DOI: 10.1080/13698230903326331.

[26] United Nations General Assembly, resolution 43/53, 1988. http://www.un.org/documents/ga/res/43/a43r053.htm (accessed November 12, 2012).

[27] See also Eric A. Posner and David Weisbach, *Climate Change Justice* (Princeton, NJ: Princeton University Press, 2010), p. 102.

[28] For a defense of a fault conception of collective responsibility, see Steve Vanderheiden, *Atmospheric Justice: A Political Theory of Climate Change* (Oxford: Oxford University Press, 2008), chp. 6.

the problem of ignorance applies equally well to states. Third, the problem of insufficient collection would also occur when assigning liability to states on the basis of their historic emissions. Most states will fail to meet the knowledge condition before the late 1980s and early 1990s; and the many new states, which were born in the twentieth century, cannot be faulted any more for emissions before their coming into existence than individual people can for emissions before their birth.[29]

The fourth problem is unique to the assignment of responsibility to states and is related to the moral presumption behind fault liability – namely, that only those at fault should be sanctioned. There are great practical barriers to assigning responsibility to states without preventing the costs from devolving to the citizens of the states because state revenues are raised by taxing citizens. In effect, then, the responsibility devolves with the costs. So, were an international institutional arrangement to lay fault on a state for all its emissions since, say, 1988, it would for practical purposes assign costs to the citizens of that state – an increasingly large proportion of whom were born after that date. These people would be assigned responsibility for emissions occurring before they were born. This is the problem of the misapplication of sanction to people not at fault. The misapplication of sanction is not a decisive reason to reject those accounts of responsibility that do not take fault as a necessary condition of responsibility, but it seems to undermine the moral basis of a fault account. An institutional arrangement ordered on the principle of fault liability that allowed costs to flow to individuals who are not liable under fault would seem to have too many leaks to be seaworthy.

Given these problems, a no-fault account of historical responsibility might seem more plausible. No-fault conceptions of responsibility simply deny at least one of the aforementioned conditions for assigning responsibility on the basis of fault. One prominent account in the law of torts or delict is strict liability, which holds agents responsible for their harmful outcomes, regardless of whether they knowingly contravened a standard of care. Given the problems with attributing violations of an appropriate standard of care or with imputing knowledge to many past agents, strict liability might seem to be a better basis for responsibility for historic emissions.[30]

---

[29] Stephen M. Gardiner suggests that the persistence of states over time solves the problem of dead emitters for an account of responsibility based on historic emissions. See *Moral Storm*, p. 418. But the picture is complicated by the existence of new states. How is South Sudan's responsibility to be calculated? Or Scotland's, should it opt for independence?

[30] In his invocation of the maxim "[i]f you broke it, you bought it," James Garvey seems to recommend a version of strict liability. See his *The Ethics of Climate Change: Right and Wrong in a Warming World* (London: Continuum, 2008), p. 75. Stephen M. Gardiner invokes a similar maxim. See *Moral Storm*, p. 416. The maxim seems most appropriate precisely when

Strict liability is sometimes criticized on grounds that it is unfair to hold persons responsible for the costs of their actions when they are not at fault.[31] In the law, strict liability is typically applied only with respect to activities that are especially dangerous or important to human health for which we seek incentives to act with a great deal of care and when agents can be put on notice beforehand that they will be held liable for the negative effects of their actions, even if they are not at fault. The brunt of the charge of unfairness might be deflected by the response that persons engaging in these dangerous kinds of activity have been put on notice that they are assuming the risks of being held responsible. That response, however, does not help in the case of climate change if the concern is to assign responsibility for historic emissions. Absent a time machine, we have no way to put people retrospectively on notice before they emit that they will be held strictly responsible at a later date.[32] Moreover, as in the application of fault liability, there is a problem of insufficient collection. A huge percentage of the people who would be strictly liable are now dead. This problem can be partially circumvented if the responsibility is assigned to states, but only if the states have a continuous existence over the time period for which responsibility is assigned. And the issue of fairness would, in any case, remain when determining which states would be held retrospectively strictly responsible.

One primary purpose of a system of strict liability is to establish incentives either to refrain from especially dangerous activity altogether or to be exceedingly cautious when engaging in it. But such incentive effects can only work if persons are on notice of the liability attached to the activity. Thus, a retrospective application of strict liability not only seems unfair, but it also

one is on notice that it is in effect, which, as I argue in the next paragraph, is true of strict liability generally. To anyone who thinks that the knowledge and standard of care conditions of fault are generally unnecessary for assigning responsibility, the history of case law in torts or delict must look utterly incomprehensible. Gardiner might be construed as supporting a principle of strict liability with his example of a person *unwittingly* eating the pizza of a colleague, after the latter dashed off without time to eat lunch. See *Ibid.* But maybe not since Gardiner is careful to call the eating "morally relevant," but does not attribute moral responsibility to the eater. The example is under described; for example, it is unclear what the norm in the office is. It might be, "don't eat other people's food." If so, the eater seems responsible under fault. It might be, "what you leave in the refrigerator is up for grabs." In that case, the eater does not seem to be morally responsible.

[31] However, Honoré argues that fault liability contains an aspect of strict liability for people who are incapable of meeting the standard of care. See Honoré, *Responsibility and Fault*, chp. 1.

[32] See also David Weisbach, "Negligence, Strict Liability, and Responsibility for Climate Change" The Harvard Project on International Climate Agreements (2010), discussion paper 10–39: 33. http://belfercenter.hks.harvard.edu/files/WeisbachDP39.pdf (accessed June 27, 2013).

defeats one of the primary purposes of strict liability, which is to establish a system of incentives to act with a great deal of care. If a system of strict liability unfairly shifts costs to people who were not put on notice, and if it provides no incentives for either not engaging in very risking courses of conduct or taking extraordinary care when doing so, there would seem to be neither a moral nor a pragmatic justification for it.

Beneficiary responsibility is also a no-fault conception of responsibility that maintains a certain connection with historical emissions, although less directly than either fault- or strict liability. Beneficiary responsibility does not require outcome responsibility, but it maintains a connection to the outcome through the relation of benefiting. The idea is that highly developed countries are assigned greater responsibility under a mitigation plan because the high standards of living that residents of these countries enjoy are the benefits of past greenhouse gas emissions. In comparison to the residents of developed countries, people now living in least-developed and developing countries have benefited much less (or not at all), and their states would be assigned a proportionally smaller mitigation burden.[33]

Does being a beneficiary provide a good moral reason for assigning a responsibility to a person? It certainly does not in general. Suppose I benefit from my neighbor painting his house because it serves to maintain the property values in my neighborhood. There is a sense in which I am free riding on his efforts and expenditure. I have done nothing; he has put himself out. Still, I enjoy a benefit.[34] But the claim that I am responsible for paying for a portion of his painting costs or directing a fraction of the proceeds of the sale of my home to him is implausible. My benefit in this case is simply the enjoyment of a positive externality of his action. That is not the kind of thing that generally obliges me to reimburse those who have acted. If it were, many of us would be in debt many times over, including to those who keep better care of their houses than we do and thereby help to maintain our property values; to the wealthy whose superior effective demand brings stores to the neighborhood that we enjoy, but could not hope to patronize enough to keep in business; to employers other than our own who keep unemployment down and desperation and crime in our cities at bay; to those much more learned than us whose contribution

---

[33] Eric Neumayer defends beneficiary responsibility in his "In Defence of Historical Accountability for Greenhouse Emissions," *Ecological Economics* 33 (2000): 185–192. Henry Shue is also sympathetic to it in "Historical Responsibility," 3.

[34] Axel Gosseries takes three conditions stated in the text to be paradigmatic of free riding, and sufficient for engendering a responsibility to compensate. See Axel Gosseries, "Historical Emissions and Free-Riding," *Ethical Perspectives* (11): 36–60. The argument of the paragraph disputes that view.

to the surrounding intellectual atmosphere enriches our lives; and even to those who spend time and money making themselves look pleasing. The list would go on and on. Because nearly everyone is a beneficiary of many positive externalities and the provider of only a few at best, the exposure to liability of most everyone living under a rule in which we were required to reimburse everyone who provided us with a positive externality would be enormous.

The argument just made casts doubt on beneficiary responsibility in general. Still, there might be other kinds of cases where benefiting is sufficient for assigning responsibility. In the previous examples, no one is benefitting from a past injustice. There is a long tradition of thought that holds that beneficiaries of injustice have a responsibility to return that which they have come to possess as a result of an unjust war. Hugo Grotius argued this point in the 17th century:

> But further, tho' a Man has not done the Damage himself, or if he did it without any Fault of his, but yet keeps in his Possession a Thing taken away by another in an unjust War, he is obliged to restore it; because there can be no Reason produced naturally just, why the other should be deprived of it.[35]

According to this account the rights of the person entitled to the object is the basis of the responsibility of the beneficiary to return the object. Beneficiary responsibility for past injustice is also often a feature of municipal law. Under United States law (18 USC § 2315), a person knowingly receiving stolen goods that have crossed borders is guilty of a federal crime. Under some state jurisdictions, one can be prosecuted for a lesser crime if one unknowingly possesses stolen goods.

The political theorist Daniel Butt argues that states, or their citizens, who benefit from historic injustices such as territorial conquest, can reasonably be held responsible for compensatory payments to persons who have been harmed by the injustice.[36] Butt contends that this is the case irrespective of whether the victims of the historic injustice are in a condition of desperate need as a result of the injustice. His argument invokes the following moral premise: "The individual's duty not to benefit from another's suffering when that suffering is a result of injustice stems from one's moral condemnation of the unjust act itself."[37] If we accept the premise, and if we condemn an unjust act, then we are under a duty not to benefit from the act. So everything turns on the premise. In support of it, Butt offers the following:

---

[35] Hugo Grotius, *The Rights of War and Peace* Bk. III, Richard Tuck ed. (Indianapolis: Liberty Fund, 2005), p. 1418.

[36] Daniel Butt, *Rectifying International Injustice: Principles of Compensation and Restitution Between Nations* (Oxford: Oxford University Press, 2009), chp. 4.

[37] *Ibid.*, p. 127.

My claim is that taking our nature as moral agents seriously requires not only that we be willing not to commit acts of injustice ourselves, but that we hold a genuine aversion to injustice and its lasting effects. We make a conceptual error if we condemn a given action as unjust, but are not willing to reverse or mitigate its effects on the grounds that it has benefited us.[38]

The reference to a conceptual error signals that Butt thinks that there is some kind of contradiction involved in denying beneficiary responsibility and condemning the original transgression.

Butt's argument does not seem to take us to his conclusion about beneficiary responsibility in cases of past injustices. The condemnation of an act as unjust does not seem to be contradicted by the claim that the beneficiaries – not the agents – of the action are not morally required to restore the victim. If receiving a benefit from an injustice were wrong, then perhaps one could not consistently condemn the original injustice and not condemn the benefit received from it. But that would just be to assume the conclusion. Another tack would be to claim that when a beneficiary (not the agent) of an injustice refuses to repay the victim, but condemns the original injustice, the beneficiary is caught in a performative contradiction.[39] But this is no better; in order for the action of refusing to repay the victim to contradict the condemnation of the original injustice, the action of refusing must be based on an implicit principle that contradicts the condemnation. But again, that's exactly the principle the argument needs to establish. Therefore, it might be wrong for beneficiaries of injustices not to repay the victims. It seems intuitively plausible in some cases at least, but there does not seem to be a conceptual error in denying it.

An alternative defense is developed by the political theorist Edward A. Page.[40] He distinguishes wrongdoing of the tort liability kind from injustice. Injustice may occur in the absence of the application of fault or strict liability standards. This might allow historical emission to be unjust even if not something for which emitters can be assigned fault or strict liability. Since prior to the end of the 1980s there was not generalized knowledge of the severe problems caused by CO_2 emissions, if it were an injustice to emit CO_2, it must be an injustice for which the emitters are not morally responsible. We may not reproach people for committing an injustice that they could not have known was an injustice. We could think of analogies involving the damaging of the

---

[38] *Ibid.*, pp. 127–128.

[39] The notion of a performative contradiction is employed by Jürgen Habermas in "Discourse Ethics: Notes on Program of Philosophical Justification" in *Moral Consciousness and Communicative Action* (Cambridge, MA: The Massachusetts Institute Press, 1991), p. 95.

[40] Edward A. Page, "Give it Up for Climate Change: A Defence of the Beneficiary Pays Principle," *International Theory* 4 (2012): 313–317.

property of another when one could not have known either that it was the property of another or that what one was doing was damage; so far, so good. Unlike in the case of fault, we can in principal envisage an historical injustice committed in ignorance.

In order to apply the idea of historical injustice to responsibility for enjoying the benefits of historical emissions, we must ask whether the historical emissions, especially those before the late 1980s, are plausible instances of historic injustice. Page notes that three accounts of historical injustice for climate change seem to get the most attention.[41] The first two resemble the injustice that I argued would occur if our generation failed to mitigate climate change. Our failure to mitigate would involve us exceeding the morally constrained cumulative emissions budget. We would thereby be emitting more than our fair share of the total allowed CO_2 and be undermining the moral choice of future generations. Page's purpose, and ours presently, is to assign such responsibility intragenerationally. The idea would have to be that previous generations had a carbon budget that was a part of the cumulative budget and that that budget was the basis for the distribution of shares among parties (either states or individuals). In the past, parties committed injustices by exceeding their allowance. Parties who now enjoy the benefits of such excess are enjoying unjust enrichment.

We have a clear and vivid sense of the upper limit of the carbon budget for our generation because at the present emissions level – within the lifetime of many people now alive – we might shoot past the cumulative budget. Doing so would certainly exceed our budget. It is far less clear what the carbon budget for any previous generation might have been if our generation had begun seriously to mitigate climate change in the 1990s, early emissions would look less problematic than they do now. It would be hard then to indentify historical generational injustice. And in that case, it would also be hard to identify any particular parties within past generations who took more than their fair share. As an account of historical injustice, this seems implausible. Whether a party has committed an injustice by emitting should not depend on whether subsequent parties sufficiently mitigate.[42] The same problem seems to afflict the third account of historical injustice for climate change noted by Page because it involves trespass by means of overusing the atmospheric commons. Again, what looks like overuse will depend on the actions of subsequent parties.

[41] *Ibid.* p. 315.

[42] David Miller makes a similar point in his "Global Justice and Climate Change: How Should Responsibilities be Distributed?" The Tanner Lecture on Human Values, Tsinghua University, Beijing, March 24–25, 2008, p. 133. http://tannerlectures.utah.edu/_documents/a-to-z/m/Miller_08.pdf (accessed June 28, 2013).

Each of the three bases for assigning responsibility for climate change mitigation on the basis of a connection to historic emissions – fault, strict liability, and beneficiary liability – contain serious, seemingly devastating, problems.[43] But, regardless of which conception of responsibility is used to support it, the polluter-pays principle and its close cousins are subject to a general problem when used for the purpose of assigning responsibility under an international mitigation regime. Even if the principle can direct the assignment of the burdens of reducing CO_2 emissions (or financing adaptation), it is silent on permission to emit CO_2 in order to fuel poverty-eradicating economic growth.[44] As emissions grow in the developing world, assigning responsibility for mitigating climate change to states on the basis of a state's total emissions would result in assigning costs increasingly to developing states. If the other problems with fault, strict liability, and beneficiary responsibility in this context were not significant, one might be tempted to think that those emissions provide the right kind of reason for laying an increasing burden of mitigation on such states. After all, in order to have a reasonably good chance of keeping warming under 2°C, we will have to limit cumulative emission to a trillion tons. To meet that goal, all states will eventually have to transform their economies away from fossil fuels. But our question is what the guiding values of such a transition should be. Surely the avoidance of danger is one, but, as I argued in Chapter 1, states have no reason to think of climate change as especially dangerous if the mitigation plan will not allow poverty eradicating economic growth. Another important value is the right to sustainable development. Taking that right seriously involves not judging a state's mitigation burden by its emissions alone, but also by whether an alternative to greenhouse gas emissions is available that will allow continued human development.

THE ABILITY-TO-PAY PRINCIPLE

Fault liability and some versions of beneficiary responsibility require the existence of an historic injustice. The problem for these accounts is in offering a convincing account that past emissions constitute an injustice. The recognition of that problem suggests an alternative. Is it plausible that being a privileged member of a certain kind of arrangement, rather than the recipient

---

[43] I have discussed these as three distinct approaches, but combinations may be possible. For an interesting combination of strict liability and beneficiary pays, see Derek Bell, "Global Climate Justice, Historic Emissions, and Excusable Ignorance," *The Monist* 94 (2011): 391–411.

[44] See Darrel Moellendorf, "Treaty Norms and Climate Change Mitigation," *Ethics and International Affairs* 23 (2009): 247–265. http://www.carnegiecouncil.org/publications/journal/233/features/001 (accessed December 19, 2012).

of benefit produced by an historic injustice, could be the reason one has a responsibility? The idea suggests moving the discussion away from considerations of tort or delict to social responsibility. John Rawls argues that the advantages that the relatively privileged person enjoys in society, conceived as a cooperative scheme for mutual benefit, are justified only if the institutions that permit such privileges also render persons in the lowest position in society better off than persons in the lowest position would be in any competing institutional order. The idea is to judge institutional orders by the condition of the least well-off who would live under them.[45] This could be construed as a particular kind of beneficiary responsibility. A person has no entitlement simply to benefit as much as possible from a scheme of cooperation; rather, one's comparative privilege entails a social responsibility to uphold an order in which the lowest positions in society are higher than in any other order. Rawls's view is meant to apply to an institutional order conceived of as a system of rules, which constitute the main institutions of a national economy. And there is considerable controversy about the merits of a similar kind of reasoning applicable to economic relations across state borders.[46] There is no hope of settling that debate here, so I set it aside.

Still, the Rawlsian idea that fairness in a cooperative scheme for mutual advantage requires that the institutions that mediate the cooperation be judged by their effects on the least well-off is relevant to an international framework regulating the energy use of states. Access to energy is tremendously advantageous.[47] It is important to states for reasons that are well understood. Typical persons living in highly developed countries require an extraordinary amount of energy to sustain their lifestyles. The incomes, relative freedom from morbidity and premature mortality, and the high-quality educations enjoyed by residents of highly human-developed states require very productive economies. And such economies require massive amounts of energy. For

---

[45] See John Rawls, *A Theory of Justice* rev. ed. (Cambridge, MA: Harvard University Press, 1999), pp. 65–69.

[46] Important representatives of the early cosmopolitan application of the Rawlsian argument are Charles Beitz, *Political Theory and International Relations* rev. ed. (Princeton, NJ: Princeton University Press, 1999) and Thomas Pogge, *Realizing Rawls* (Ithaca, NY: Cornell University Press, 1989), pt. 3. Rawls responds in John Rawls, *The Law of Peoples* (Cambridge, MA: Harvard University Press, 1999). The debate in the literature continues. See also my *Cosmopolitan Justice* (Boulder: Westview Press, 2002).

[47] My claim is limited here to the invocation of a line of reasoning drawn from Rawls's defense of the difference principle. For an argument against the suitability of Rawls's philosophy to understanding our obligations in the face of a global environmental threat, see Stephen M. Gardiner, "Rawls and Climate Change: Does Rawlsian Political Philosophy Pass the Global Test?" *Critical Review of International Social and Political Philosophy* 14 (2001): 125–151.

laudable reasons, then, states seek to provide inexpensive, widely available energy. The threat of dangerous climate change, however, requires a large-scale transition away from the use of inexpensive fossil fuels. This transition will necessarily impose costs on some parties, as measured against their current expenditures on energy, but importantly probably not on parties as measured against their future prospects under a 4°C warming scenario. There is much uncertainty that surrounds climate change forecasts, but it is not unreasonable to suppose that all states have reasons to want to prevent a world that is 4°C warmer. An international agreement that prevented such a scenario, therefore, would be a cooperative arrangement to the mutual advantage of all parties.

An international climate change agreement will also affect access to inexpensive energy. In light of the moral importance of human development, in order for the costs of the transition to a global no-carbon economy to be fair, they must not slow poverty-eradicating human development. A fair international framework of access to energy will ensure that the prospects of poor states to pursue human development are not made worse than under a business-as-usual scenario. The poor living in least-developed and developing states could reasonably reject any principle for intragenerational assignment of responsibility for climate change mitigation that rendered the human development prospects of their states worse off than under a business-as-usual scenario. This idea lends credence to the view that an architecture of responsibility for climate change mitigation should be based on states' ability to pay.

It might be objected that states that come to an international regime already relatively rich should not have to sacrifice under that new regime even if it is for the laudable goal of overcoming poverty in states that come to the regime with lower levels of human development. Pre-existing privileges should not be the basis of responsibilities that require sacrifices under a new cooperative arrangement.[48] Whether or not that claim is reasonable, it is not accurate in application to the present case. It is reasonable to believe that all states can benefit from an international arrangement that prevents warming of 4°C. States that come to an international regime already rich are not making sacrifices if it is reasonable to believe that they are benefitting against the business – as – usual baseline. The Rawlsian point is that they have no entitlement to benefit as much as possible from that agreement. Not hindering poverty eradication in least developed and developing countries constrains the benefits that already wealthy states may expect from a climate change agreement.

---

[48] I am thankful to Henry S. Richardson for discussions on matters related this and the surrounding paragraphs.

The ability-to-pay principle is a third kind of no-fault principle. It assigns responsibility in proportion to an agent's capacity – variously measured. Ability-to-pay and closely related no-fault conceptions of responsibility are often used in the assignment of responsibility for financing state activities, such as defense against various threats and the provision of certain aspects of the well-being of citizens. Generally, in financing programs directed to meet these aims, states do not look for citizens who are at fault for the wealth that they possess. Progressive income taxation to raise public revenue for the provision of goods and services is typically defended on grounds that the wealthier have a greater ability to pay. The assignment of responsibility for climate change under an international climate change regime is analogous to the responsibility under a municipal tax regime.

The proposal is to assign states responsibility for mitigation in accordance with their ability to pay measured broadly in terms of their level of human development. This would protect states with low levels of human development from assuming mitigation burdens that might delay poverty eradication. The social pressures propelling human development are probably strong enough to allay worries that such a system might discourage human development by laying a burden on states that are successful in advancing human development. And proper design of the system can further allay such worries.[49] If burdens are assigned within ranges of human development and not for each marginal increase, any disincentive would apply only to a marginal improvement in human development that moved a state from one broad range to another. If the ranges are wide, a state could be assured of further prospects for human development without additional burdens until moving into the next range. The disincentive of crossing the threshold from one range to the next can be overcome by the prospect of further progress in human development within the range without the penalty of additional mitigation burdens.

The proposal is very general in several ways. First, implementing the proposal would obviously require making judgments about the level of human development that is the threshold according to which mitigation burdens are assigned differentially to states in the range below and above the line. The idea is to fulfill the general goal of assigning responsibility in a way that is consistent with the right to sustainable development. Particular assignments may contain some degree of arbitrariness. The difference in levels of human development of states just above and just below a threshold separating ranges may be small. The only reasonable justification of the differential treatment

---

[49] I am thankful to Klaus Günther and Julian Culp for helping me to see the importance of these matters.

that would follow from a distinction like that is that it fulfills the more general aim of preserving the capacity of states to pursue human development because it is highly unlikely that there would be a significant difference between the states that would otherwise merit the difference in treatment. However, the proposal allows for flexibility in setting the thresholds.

Second, this general proposal is silent on what assuming additional responsibilities for mitigation would require. For example, it could require either domestic per capita emissions reductions or subsidizing the purchase of renewable energy in states that have lower levels of human development. Third, the time period during which states with lower levels of human development may assume lighter mitigation burdens is not specified. There is a physical limit to our greenhouse gas emissions if there is to be reasonable hope of hitting the goal of limiting warming to 2°C or any other warming limit. Any realistic assignment of mitigation responsibilities must be brought into line with a temperature target.

In contrast to the polluter-pays principle, the ability-to-pay principle fits particularly well within the project of establishing responsibilities under an international framework intended to regulate energy use and permit human development. Fault and strict liability are best fitted to assigning responsibility for deviations against a stable background of entitlements. They are not well suited to assigning that background of entitlements. Here considerations of fairness are particularly important. Although polluting activity that is contrary to municipal law may be handled reasonably well by criminal or tort law, the determination of fair access to energy and entitlements to the use of various forms of energy among states, when the well-being of billions of people is at stake, amounts to establishing or reforming some of the background rules of relations between states.

OBJECTIONS AND REPLIES

One criticism of developing the background architecture of social responsibility in an international climate regime solely on the basis of the ability-to-pay principle is that it is counterintuitive to disregard the historical causes of the problem.[50] The philosopher Simon Caney claims that, in addition to being most able to pay, part of what makes wealthy parties legitimate targets for the assignment of responsibility is that "the wealth they hold came about in

---

[50] Simon Caney, "Human Rights, Responsibilities, and Climate Change," in Charles R. Beitz and Robert E. Goodin, eds., *Global Basic Rights* (Oxford: Oxford University Press, 2009), p. 242. See also his "Climate Change and the Duties of the Advantaged," pp. 214–216 and 217–218.

climate-endangering or other unjust ways."[51] Because Caney acknowledges that no current highly developed countries developed without massive CO_2 emissions, the point of the addition is not to pick out a different group of parties responsible to pay, but to avoid the otherwise alleged counterintuitive character of targeting only those able to pay.

As I argued near the beginning of this chapter, conceptions of responsibility will vary appropriately according to their purpose. If our purpose in assigning responsibility is to sanction people who have acquired their wealth unjustly or unfairly, then of course the historical origins of their wealth matter. This is not the appropriate purpose for the assignment of responsibility within an international framework regulating access to energy in order to achieve climate change mitigation. Here the overriding purposes are the achievement of the mitigation objective and the safeguarding of human development. The former requires large global reductions in emissions; the latter requires continued access to inexpensive energy in the least-developed and developing countries. A schedule of emission reductions that assigns greater burdens to the highly human-developed states and fewer to less-developed states, and the requirement to subsidize the transition to alternative energy forms in these countries can be justified by the greater ability of the highly human developed states to pay. Ability to pay is a sufficient justification for assigning burdens in this way because it serves the important moral purposes of the framework to avert dangerous climate change and to provide fair access to energy.

In the context of a framework for state interaction, the relevant abilities to pay are those of states. As the previous discussion showed, one objection to the assignment of responsibility to states rather than to individuals is that responsibility would unfairly devolve to individuals. But the force of the objection is in an important way less strong when applied to no-fault conceptions of responsibility. According to no-fault conceptions of responsibility, if a responsibility devolves to citizens to pay for a state obligation, no stigma of fault devolves with it.

There may, however, be other problems of fairness. A consequence of assigning responsibility to states on the basis of ability to pay is that wealthy people in poor states might carry a lighter burden than less-wealthy people in richer states simply because the wealthy people live in a poor state.[52] This raises two moral challenges. First, the manner in which the responsibility for mitigation might

---

[51] Caney, "Human Rights, Responsibilities, and Climate Change," p. 244.

[52] Paul Harris objects to this consequence in his *World Ethics and Climate Change: From International Justice to Global Justice* (Edinburgh: University of Edinburgh Press, 2010), pp. 131–132 and 143–152. Simon Caney also objects to it in his "Human Rights, Responsibilities, and Climate Change," pp. 245–246.

devolve to individuals could be in tension with justification of the principle to assign responsibility according to ability to pay. Second, regardless of any internal tensions in the account, perhaps it is morally objectionable on good independent grounds to have a framework for cooperation among states in which some less-wealthy people are more burdened than some more-wealthy ones.

The first challenge is, I think, easily handled. If the point of assigning responsibility for climate change is to reduce inequalities between persons, then an institutional order that burdened some less-wealthy persons more than some more-wealthy ones is inconsistent with the aim. But if the conception of responsibility is based on the claim of developing states to be able to pursue human development, then there is no contradiction. Human development is valuable because it involves the production of valuable goods – in particular, education, health, and income. Highly developed states might nonetheless be very unequal, as is the case with the United States of America. The aim of securing human development is not necessarily egalitarian. Insofar as that is the case, there is no contradiction internal to an account that aims at promoting development but does not pursue egalitarian objectives.

The rebuttal to the first challenge probably increases the apparent force of the second. Perhaps merely securing human development is an inadequate goal, especially if doing so would permit exacerbating inequalities between rich people in poor countries and the non-rich in rich countries. There is, I would agree, good reason to think that a more egalitarian world order would be more just than the present world order,[53] but that does not settle the question of what the aims of a climate change agreement should be. An institutional order comprises multiple institutions directed toward various goals, not all of which need to be directed toward the realization of egalitarian justice in order to be morally appropriate. This is the case even if the complete order – the interworking of the various institutions – should satisfy egalitarian justice.

The problem identified in the second challenge is not peculiar to an international climate change agreement. Presumably, we expect an educational system to provide education in a way that contributes to the goal of equality of opportunity. No educational system alone, however, can realize that goal because the families of children will have different resources and these will be available to children in the pursuit of their opportunities. Addressing this by a policy of providing more resources to schools in poorer neighborhoods looks promising. But not all of the people who live in such neighborhoods are poor. The policy would inadvertently benefit some children from more

---

[53] See Darrel Moellendorf, *Global Inequality Matters* (Basingstoke: Palgrave Macmillan, 2009).

well-off families more than poor children not living in poor neighborhoods. But through only the means of educational policy, the tools that we have to work with are not suited to solving that kind of a problem.

The limitations on climate change policy are similar in some ways. States with high levels of human development typically have a class of poor people, but generally such states do not contain widespread extreme poverty. Some states with lower human development are home to a small class of extraordinarily rich people. Through securing the right to human development we do much good, but we cannot ensure that everyone gets what they should have (according to egalitarian justice). The ambitions of an international climate change treaty should be considerably lower than that.

The architecture of responsibility for a mitigation regime should employ a conception of social responsibility, rather than the remedial responsibility of the law of torts. The most reasonable conception for that purpose would assign responsibility to states based on the ability to pay, where such ability is understood broadly in terms of the state's level of human development. This will not ensure that all and only wealthy people pay for climate change mitigation, but we do not generally hold a policy hostage to that requirement even when, as is the case with education policy, the policy is rightly directed toward egalitarian aims.

7

Urgency and Policy

"Human progress is neither automatic nor inevitable. We are faced now with the fact that tomorrow is today. We are confronted with the fierce urgency of now. In this unfolding conundrum of life and history there is such a thing as being too late . . . We may cry out desperately for time to pause in her passage, but time is deaf to every plea and rushes on. Over the bleached bones and jumbled residues of numerous civilizations are written the pathetic words: Too late."

– Martin Luther King, Jr.

The warming limit of 2°C has been widely endorsed. The European Union and several prominent international NGOs, including Oxfam and Christian Aid, had advocated the limit for several years before the sixteenth Conference of the Parties (COP 16) to the UNFCCC formally adopted it in Cancun in 2010.[1] The temperature goal is, however, to a large extent arbitrary. Because of cascading uncertainties associated with various feedbacks in the climate system, we are unable to predict if dire consequences owing to warming might befall us well before an equilibrium warming of 2°C. Moreover, whether it is reasonable for least-developed and developing countries to accept a mitigation plan designed to keep warming within that temperature target depends on whether the plan lays costs on them that are likely to prolong or deepen poverty by raising their energy costs. But because it seems possible to limit the warming to 2°C

---

[1] See UNFCCC, *Report of the Conference of the Parties on Its Sixteenth Session, Held in Cancun from 29 November to 10 December 2010.* http://unfccc.int/resource/docs/2010/cop16/eng/07a01 .pdf#page=2 (accessed November 22, 2012). The Copenhagen Accord produced by COP 15 also affirmed the 2°C limit, but the status of the Accord as merely noted, rather than affirmed, left the commitment to the limit somewhat uncertain. See UNFCCC, *Report of the Conference of the Parties on Its Fifteenth Session, Held in Copenhagen from 7 to 19 December 2009.* http://unfccc.int/resource/docs/2009/cop15/eng/11a01.pdf (accessed November 22, 2012).

in a manner that would prevent heaping costs on the least-developed and developing countries, and because of the projected economic costs of trying to limit warming at a lower temperature, the 2°C limit has broad credibility. Despite this credibility, international negotiations have so far failed to produce an agreement that would limit emissions of greenhouse gases sufficiently to have a good chance of achieving the warming limit.

I begin this chapter by explaining the urgency if the 2°C warming limit is to remain a real possibility for international negotiations. Obviously, it would be best if appreciation of this situation produced greater commitment than currently exists among the major emitting states (especially the United States) to take leadership in producing an effective international mitigation regime. Although we may hope for that, and while we should attempt to hold political officeholders accountable, the increasing chances of failure at limiting warming to 2°C raises the important question of what, if anything, we should do to plan for a future planet that is warmer than 2°C.

Next I discuss an important argument that we should not invest research and resources into such planning. This seems wrong to me, and I explain why I think so. I then discuss and evaluate three areas of policy: increased planning and investment in adaptation to protect people; studies and trials in assisting the colonization or migration of plant and animal species into new areas to promote their survival; research into climate- or geo-engineering proposals to control the effects of CO_2 emissions so as to reduce warming in the absence of emission reductions that would otherwise be required to limit warming to 2°C. Although each of these may be an important supplement to serious mitigation, none is a substitute.

In the final sections of this chapter I return to our best hope for avoiding dangerous climate change: an effective international agreement on mitigation. I discuss the conditions that such an agreement would have to satisfy in order to be morally satisfactory as well as a policy – known as *pledge and review* – that aims to achieve mitigation. The Copenhagen Accord (the Accord) adopted at COP 15 in 2009 includes a very weak and inadequate version of pledge and review. More sophisticated versions of pledge and review should be capable of solving some of the moral problems of the Accord's version. Still, a central worry about all such proposals is that they may simply be too little, too late.

THE FIERCE URGENCY OF NOW

I have discussed the morally constrained CO_2 emissions budget in previous chapters. For now I will just underscore the urgency of the constraints. Recent scientific studies of the relationship between warming and CO_2 emissions emphasize the importance of cumulative anthropogenic emissions. From the industrial revolution until the year 2000, about a half-a-trillion tons of carbon were put into the atmosphere by humans. In order to have a good chance of limiting warming to $2^{\circ}C$, these studies indicate that total carbon emissions must not exceed 1 trillion tons. But as of this writing, total emissions continue to increase; we are on a trajectory to emit the next half-a-trillion tons by 2040 or sooner. If we do that, we will have surely exceeded our share of the CO_2 emissions budget because humanity's total budget will have been consumed.

The budget can be extended only by reducing emissions, not by raising the ceiling. The sooner – and therefore lower – emissions peak, the less precipitously they must subsequently fall. The longer we wait to begin reducing emissions, the steeper the reduction curve must be not to exceed the budget. As suggested earlier, a serious risk of delaying mitigation is that abrupt reductions taken later could produce a global economic recession. Therefore, the failure of negotiators to achieve an emissions reduction regime in Copenhagen, as had been hoped, and the subsequent postponement of the deadline for implementing a plan until 2020 increases the possibility that a plan to make the necessary reductions will require unacceptably high economic costs.

Perhaps waiting would allow for the possibility of technological breakthroughs in renewable energy that would substantially reduce the costs of the transition or, failing that, the costs of transitioning to a no-carbon economy would be easier to bear as a result of economic growth. We should not assume that a breakthrough is impossible, but the mere hope of one hardly amounts to a basis for responsible policy. The needed technological breakthroughs are far more likely to come about if a mitigation plan raising the costs of fossil fuels is in place because it would provide incentives to consume alternative fuels. Stronger market demand for alternative fuels gives rise to competition among producers, thereby increasing the likelihood of technological advances. And although it is generally true that costs relative to income go down as income rises, this is the case only if the costs remain the same. The longer we wait to begin reducing emissions, the more sharply our emissions will have to fall, which means that we will have to rely more heavily on renewable technologies and we may have to abandon costs that in the meantime are sunk into fossil fuel production and consumption.

There are powerful interests resisting the urgency of mitigating climate change. Fossil fuels are big business. There will be a strong push to develop and exploit known and suspected underground reserves, which contain up to 2,860 billion tons (Gt) of CO_2, significantly exceeding the morally constrained CO_2 emissions limit.[2] And there will be significant costs sunk into the capital investments of coal-fired electricity generation plants built now or in the near future. We are about to witness an explosion in the production of such plants in Asia.[3] This makes for powerful lobbies and alliances against establishing a price on CO_2 emissions.

MORAL HAZARD

It is unclear whether we are going to have an effective international mitigation agreement in time to stay on budget. Recognition of the possibility of failure in that regard is the motivation for considering the merits of alternatives, including increased planning and investment in adaptation to protect people, trials of assisted migration to protect some plant and animal species, and research into geo-engineering to protect against catastrophic climate change. There is, however, a worry that engaging in research and planning in these areas will exacerbate the problem of implementing effective mitigation. Insurers know well the problem that certain forms of insurance coverage might dispel a reason that a person otherwise would have not to engage in risky behavior. A person who did not snowboard might wait until getting health insurance before pursuing the hobby; the existence of insurance can make some risky conduct more attractive. This is the problem of the moral hazard.

Perhaps activities such as planning more effective means of adaptation for humans and their communities, helping species colonize new geographic areas, and researching geo-engineering are like taking out an insurance policy that removes a reason not to engage in risky behavior by offering protection against a potential mishap.[4] If being better prepared to adapt to climate change, knowing how to preserve more species, and having the means to geo-engineer the planet each weakened our resolve to reduce greenhouse gas emissions, then perhaps they constitute moral hazards that policy should seek to avoid.

---

[2] Carbon Tracker, "Unburnable Carbon 2013: Wasted Capital and Stranded Assets." http://www.carbontracker.org/wp-content/uploads/downloads/2013/04/Unburnable-Carbon-2-Web-Version.pdf (accessed July 3, 2013).

[3] U.S. Energy Information Administration, "International Energy Outlook 2010 – Highlights." http://www.eia.doe.gov/oiaf/ieo/highlights.html (accessed November 1, 2012).

[4] One early expression of this worry with respect to geo-engineering can be found in Stephen H. Schneider, "Geoengineering: Could – or Should – We Do It?" *Climatic Change* 33 (1996): 296.

In general, the force of this kind of moral hazard objection is limited. The principle behind the objection is that insurance should protect only against outcomes whose likelihood of occurrence is insensitive to the existence of insurance. But that is implausible. If there are activities that would be especially valuable to us but only if we could reduce the risks associated with them, then an insurance plan that reduces those risks would be a desirable means of reducing the risks of the valuable activity. If the vast majority of snowboarders do not have debilitating and costly accidents, medical and accident insurance would be good for a person who would like to snowboard but could not otherwise afford the hospital bill in the unlikely event of a serious accident. That someone would not snowboard without health insurance is a reason to provide the insurance, not to make it unavailable.

If there is a reason to worry about moral hazards, it must be in cases where the activity is absolutely dangerous – in the sense that it ought to be avoided in all circumstances – and therefore, we should be careful not to remove a disincentive to its pursuit. We would not want business insurance to provide an incentive for businesses to neglect their premises adequately out of knowledge that they will be covered for the cost of an accident if they are negligent. One reason for a deductible that the person covered by insurance must pay is to reduce the moral hazard of the insurance. My argument in Chapter 1 suggests that it is not obvious that emitting CO_2 is necessarily dangerous in a way roughly analogous to the negligence of the business owner, although it might be. CO_2 emissions are the side effects of the important activity of generating energy at a low cost. This activity has the potential to help billions of people free themselves and their children from poverty. We should want to reduce the risk of that activity even as we seek to encourage alternatives without the side effects.

The moral hazard objection reappears, however, by asserting that the choice of the safest policy – namely, serious mitigation – is made less likely by research into practices that might make it seem less urgent. The response to this version of the moral hazard argument involves stressing the epistemic uncertainty regarding the possibility of climate change–caused catastrophes at any level of warming and the moral uncertainty concerning our commitment to serious mitigation. If we knew that the risks of very bad outcomes would only become significant after a warming of 2°C, and if we were confident that we would adopt a policy of limiting warming to 2°C, then the moral hazard argument would provide a good reason not to engage in activities that made mitigation seem less urgent. In the absence of such knowledge, the concern about very bad outcomes, even with a mitigation policy, overrides worries that a serious mitigation effort might somehow be undermined.

INCREASED ADAPTATION

One problem with current adaptation planning, such as it is, is that it is almost certainly inadequate in the event of warming in excess of 2°C. The Copenhagen Accord pledged that developed states would provide $100 billion annually beginning in 2020 to fund climate change adaptation in developing and least-developed countries. This pledge was given an institutional home at COP 16 in Cancun. The document agreed on at that meeting promised a Green Climate Fund under the trusteeship of the World Bank. But the funding sources and mechanisms still remain vague. Estimates of the costs of adaptation to climate change vary considerably and depend, among other things, on the level of expected warming. The $100 billion sum might be based on a realistic assessment of the costs of adaptation at 2°C. The economist Nicholas Stern endorses the United Nations Human Development Programme's estimates that adaptation to climate will cost about $86 billion annually if climate change mitigation policies manage to keep warming below 2–3°C.[5] But according to the UNFCCC's estimates of costs in the range of $49–$171 million, the UNDP figure is rather low.[6] And these figures are, in any case, merely to insure that the scourge of poverty is not worsened by climate change. The costs of adapting to climate change at 4°C would be significantly higher than at 2°C. According to one recent projection, just the maintenance of sea-level defenses – including dykes, walls, and gates of various sorts – would cost $270 billion annually by 2100.[7]

Regardless of whether we have to adapt to a world that is 2°C warmer or one even warmer still, how should adaptation costs be borne? It seems unfair that each state alone should be responsible for the damages within its borders.[8] This conception of responsibility would assign very high costs to poor states afflicted by damage caused by climate change. The result would be to threaten the capacity of some states to achieve significant human development. The right to human development is, then, incompatible with a conception of responsibility for adaptation that makes states entirely responsible for the problems within their borders.

[5] Nicholas Stern, *The Global Deal: Climate Change and the Creation of a New Era of Progress and Prosperity* (New York: Public Affairs, 2009), p. 70.

[6] See UNFCCC, "Investment and Financial Flows to Address Climate Change, Executive Summary," 2007. http://unfccc.int/files/cooperation_and_support/financial_mechanism/application/pdf/executive_summary.pdf (accessed December 19, 2012).

[7] Mark New, et al., "Four Degrees and Beyond: The Potential for Global Temperature Increase of Four Degrees and Its Implications," *Philosophical Transactions of the Royal Society A* 369 (2011): 11.

[8] See also Henry Shue's seminal article, "Subsistence Emissions and Luxury Emissions," *Law & Policy* 15 (1993): 39–60. DOI: 10.1111/j.1467-9930.1993.tb00093.x.

Some people favor a version of the polluter-pays principle that would seek funding from parties in proportion to their contribution to the problem and deliver the funding to affected states with the aim of making those affected whole. The responsibility of a particular state would be the result of subtracting a state's claims for damages from its share in the production of total damages.[9] If the latter is greater than the former, the state is a net debtor and must provide funds for restitution. But if the damages exceed the contribution to the problem, then the state is a net creditor and has a claim on the fund for restitution. The moral problems with this approach are familiar in light of the discussion in Chapter 6. It is not the case, in a great many instances, that people were, and still are, at fault for their emissions; strict liability retroactively applied would be unfair; and beneficiary responsibility seems implausible.

In contrast to the polluter-pays principle, the virtues of the ability-to-pay principle in this context are the same ones that it possesses for climate change in general. The principle is especially plausible with respect to establishing background institutions that set up entitlements against which fault for noncompliance can then be attributed; the Convention contains language that is well interpreted as supporting the ability-to-pay principle, and, finally, the ability-to-pay principle coheres nicely with the right to sustainable development. Developed states have the responsibility to ensure that the costs of adapting to climate change in developing and least-developed states do not undermine their human development objectives. Unlike the fault-based conceptions of responsibility, with ability-to-pay there is no prior condition to aim at in order to make the affected states whole. Therefore, there is room for reasonable disagreement about what would count as sufficient subsidies to ensure that development is not undermined. However, it is manifestly clear that many poor states have very little coping power for the climate change–related disasters that are likely to befall them, and this conception of responsibility orients action toward ensuring that they are protected.

Another problem for adaptation planning is the uncertainty inherent in all climate change forecasts. As climate scientists come to understand the climate system better, some matters will be transformed from cases of epistemic uncertainty to risks, but generally we should expect uncertainty to increase. There are several reasons for this. First, as computing power increases our ability to model more aspects of the climate system, we should expect more areas of uncertainty to emerge. Second, predictions about climate change are based

---

9 Paul Baer "Adaptation: Who Pays Whom?" in W. Adger, et al., eds., *Fairness in Adaptation to Climate Change* (Cambridge, MA: The Massachusetts Institute of Technology Press, 2006), pp. 131–156.

on cases where many computer models yield substantially similar forecasts, but because these models differ across multiple inputs and parameters, we are uncertain about a great many other matters even as we gain confidence in certain forecasts. And third, climate outcomes are not merely the result of physical processes, but also the result of the kinds and patterns of economic and population growth, about which there is moral uncertainty.[10] Given the uncertainty, it is not possible to plan for optimal investments to hedge against climate change risks. One way to respond is to focus instead on policies that are robust across a multitude of possible outcomes.[11]

Insofar as the severity of possible outcomes at greater than 2°C warming would overwhelm planning appropriate for 2°C, there is a case to be made for planning for much worse outcomes.[12] It may be necessary to move people rather than to build dykes appropriate to sea-level rise at 2°C, but as we discussed in Chapter 3, to take an outcome of climate change as one that we should protect people against, we need more than its sheer possibility. We need strong reasons to avoid the outcome, comparatively weak reasons to avoid the costs of protection, and good reasons to believe that causal antecedents of the outcome are in place. Robustness, however, can be to a greater or lesser extent, and not all outcomes can be guarded against. Decisions will have to be made about which outcomes merit advanced spending on adaption. When the pattern of outcomes is uncertain, it might make sense to await more knowledge before investing if our reasons to guard against the outcome are weaker and the costs are higher.[13]

To make matters worse, it is possible that this brief discussion of adaptation to warming beyond 2°C is based on a false premise; namely, that adaptation to a much warmer planet is uniformly possible given sufficient knowledge and money. This simply might not be the case. The interactions of outcomes at high levels of warming have not yet been very thoroughly considered. One first attempt at doing so concludes that

---

[10] The sources of possible increased uncertainty are explained in Surajae Dessai, et al., "Climate Prediction: A Limit to Adaptation," in W. Neil Adger, et al., *Adapting to Climate Change: Thresholds, Values, and Governance* (Cambridge: Cambridge University Press, 2009), pp. 68–69. Dale Jamieson also notes the possibility for an increased uncertainty with more knowledge in his seminal "Managing the Future: Public Policy, Scientific Uncertainty, and Global Warming" in Donald Scherrer, ed., *Upstream/downstream: Issue in Environmental Ethics* (Philadelphia, PA: Temple University Press, 1990), p. 85.

[11] Dessai, et al., "Climate Prediction," p. 73.

[12] See Mark Stafford Smith, et al., "Rethinking Adaptation for a 4°C World," *Philosophical Transactions of the Royal Society* A 369 (2011): 206–208.

[13] That the case of indeterminate outcomes requiring different responses might merit delayed investment is discussed in *Ibid.*, p. 209.

In . . . a 4°C world, the limits for human adaptation are likely to be exceeded in many parts of the of the world, while the limits of adaptation for natural resource systems would largely be exceeded throughout the world. Hence, the ecosystem services upon which human livelihoods depend would not be preserved. Even though some studies have suggested that adaptation in some areas might still be feasible for human systems, such assessments have generally not taken into account lost ecosystem services. Climate change impacts, especially drought and sea level rise, are likely to lead to human migration as people attempt to seek livelihoods elsewhere.[14]

The result of warming significantly beyond 2°C could be that large areas of the Earth are no longer inhabitable by humans. This is a reason to believe that, if warming increases significantly beyond 2°C, increased investment into adaptation will be of little help to a great many people. Costs will swamp policy efforts, even as the flood waters swamp the people.

ASSISTED MIGRATION OF SPECIES

The Intergovernmental Panel on Climate Change warns of the extinction of 40–70 percent of assessed species if warming exceeds 3.5°C.[15] As noted in Chapter 2, one recent study finds that species are moving to higher altitudes at the rate of 11 m per decade and toward the poles at 16.9 km per decade.[16] Not all species on the move will successfully make it. Some will find the way blocked by human communities. Others dependent on a form of mutualism will not survive in the absence of a partner species. Others still are just too far away from an ecological niche in which they might thrive. In light of the threat of mass extinction, scientific and lay interest in assisting the migration of endangered species to more hospitable regions is growing. The claims of the economic and aesthetic value of species in Chapter 2 provide at least an initial justification of assisted migration.[17]

---

[14] Rachel Warren, "The Role of Interactions in a World Implementing Adaptation and Mitigation Solutions to Climate Change," *Philosophical Transactions of the Royal Society* A 369 (2001): 234.

[15] IPCCC, 2007 *Synthesis Report*, sec. 3.2.4.

[16] See I-Ching Chen, Jane K. Hill, Ralf Ohlemüller, David B. Roy, and Chris D. Thomas, "Rapid Range Shifts of Species Associated with High Levels of Climate Warming," *Science* 333 (2011): 1024–1026. http://www.sciencemag.org/content/333/6045/1024 (accessed March 6, 2012).

[17] For an argument that rejects such an initial justification of assisted migration, see Ronald Sandler, "The Value of Species and the Ethical Foundation of Assisted Colonization," *Conservation Biology* 24 (2010): 424–431. DOI: 10.1111/j.1523–1739.2009.01351.x. Sandler and I disagree about the economic value of species and he does not consider their aesthetic value.

Whether the practice is all things considered justified depends on the strength of certain objections to it. Three objections seem strongest. One objection might be called *the green objection*. It holds that assisted migration is problematic because an important aspect of the value of species is that they are the result of natural processes, and are not human artifacts. Assisting species in their survival by means of moving them would leave the subsequent history of the saved species with a human stamp. The second objection is based on the uncertain effects that moving a species might have on the ecosystem to which it is moved. I call this *the argument from uncertainty*. The third objection can be called *the ecological dilemma*. It raises the worry that assisted migration involves the disruption of ecosystems and replaces a stability-oriented conservationist and restorationist approach to ecosystem integrity with an open-ended commitment to species preservation.

The green objection holds that an important part of the value of a species is its natural history; namely, that it is the result of natural processes of adaptation to an environment "untouched by human hands."[18] When a species is relocated by humans to a new area, the history of the species is permanently changed by human action. This change produces a loss of value. In the process of saving a species, a significant aspect of what is valuable about it is then lost.

There are, I think, three responses to this objection that together significantly undermine its strength. The first response is that in the context of the present discussion in which a species is threatened with extinction caused by anthropogenic forcing in the anthropocene, it is doubtful that the species is in any case untouched by humans. That a species would have more value if untouched by human hands is a weak reason not to assist its survival since its peril is also the result of human activity. The second response is that the objection seems too strong because it would seem to reject not only assisted migration, but all manner of species preservation, whether in situ by regulating human access to regions which species inhabit or ex situ by external preservation and eventual reintroduction. In all of these cases, insofar as the species owes its survival to the conservation practice, it is no longer untouched by humans. Yet activities of this sort to preserve species within recent historical ranges have long been practiced and are sanctioned under the U.S. Endangered Species Act.[19] The green objection condemns too much of existing preservation practice – to which we are deeply committed – to be plausible. Finally, suppose it is the case that a species would be more valuable if

---

[18] See Robert E. Goodin, *Green Political Theory* (London: Polity Press, 1992), p. 41.
[19] See Mark W. Schawrtz, et al., "Managed Relocation: Integrating the Scientific, Regulatory, and Ethical Challenges," *BioScience* 62 (2012): 737.

untouched by human hands. If (without assisted migration) its survival is in doubt, then the entirety of its value is threatened, not merely the part that derives from it being natural. It would seem better to preserve at least some of its value, say its economic value or the aesthetic value of its members that is not destroyed by being touched by human activity, than to lose all of its value.

The argument from uncertainty holds that assisted migration is to be rejected because the effects of the introduction of a species into a new region are unknown and might be negative in relation to existing species and the ecosystem generally. Application of the minimax rule (as discussed in Chapter 3) would require us to compare the extinction of the species without assisted migration with the major disruption that it could cause in the recipient ecosystem. Where that disruption includes threatening the existence of other species, there would normally be good reasons to reject assisted migration, barring some reason to value the species to be aided by assisted migration more highly than those already in the ecosystem. This places the burden on the advocate of assisted migration to argue that the most worrisome effects of transferring the species are not uncertain, but are instead unlikely. Sober assessments of assisted migration accept that there is such a burden in each case and seek research and modeling tools to overcome the uncertainty.[20]

The ecological dilemma is the worry that the results of successful assisted migration efforts – even those proceeding only after careful research and modeling – might be the preservation of many species, but the destruction of the integrity of several ecosystems. If, following Aldo Leopold's land ethic (discussed in Chapter 2), there is aesthetic value to ecosystems, and if part of the value derives from their natural origin, then that value is diminished by ecosystem adaptation to humanly introduced species. In Chapter 2 I argued that the land ethic can be understood as a claim about the aesthetic value of ecosystems and their constituents. According to the ecological dilemma, we might preserve the economic and aesthetic value of a species at the cost of diminishing the aesthetic value of the ecosystems. If – as seems to be the case – rarity increases the aesthetic value of things, as ecosystems that have not

---

[20] There is an emerging literature on these issues including: Jason S. McLachlan, et al., "A Framework for Debate of Assisted Migration in an Era of Climate Change," *Conservation Biology* 21 (2007): 297–302, DOI: 10.111/j.1523–1739.2007.00676.x, http://onlinelibrary.wiley .com/doi/10.1111/j.1523-1739.2007.00676.x/pdf (accessed December 20, 2012); Ben A. Minteer and James P. Collins, "Move It or Lose It? The Ecological Ethics of Relocating Species under Climate Change," *Ecological Applications* 20 (2010): 1801–1804, http://www.esajournals.org/ doi/pdf/10.1890/10-0318.1 (accessed December 20, 2012); Alejandro E. Camacho, "Assisted Migration: Redefining Nature and Natural Resource Law under Climate Change," *Yale Journal on Regulation* 27 (2010): 171–255; and Schwartz, et al., "Managed Relocation."

been affected by assisted migration become rarer, the marginal loss of value associated with a transplant to an unaffected area increases.

If the dilemma is real, it alone is not a sufficient reason to oppose assisted migration. Without an assessment of where the loss of value is greatest, nothing obviously follows about merits of assisted migration. The dilemma, however, seems to be overstated. A great many species are already migrating in response to global warming; there is every reason to believe that this process will continue. This migration will disrupt the equilibrium of many ecosystems. These disruptions are the result of human activity. So, anthropogenic warming on a global scale is already a threat to the integrity of ecosystems. As the disruptions grow, it will become less and less plausible to claim that assisted migration disrupts the natural integrity of ecosystems.

The initial justification of assisted migration can now be revised in light of our consideration of the three objections discussed. Neither the green objection nor the ecological dilemma establishes sufficient reason not to engage in appropriate tests of assisted migration. The argument from uncertainty, however, counsels caution and thorough research. Invasive species can disrupt ecosystems and reduce biodiversity.[21] Assisted migration may be permissible, but it should be undertaken only after good scientific research provides a credible reason to believe that the likelihood of native species being endangered by immigrant species is sufficiently low.[22]

GEO-ENGINEERING

Geo-engineering involves deliberately affecting the climate for the purpose of reducing anthropogenic warming, by means other than reducing the use of greenhouse gases.[23] Geo-engineering comprises two main kinds of technology. One is called *solar radiation management* and it seeks to increase the Earth's albedo – its capacity to reflect solar radiation back into space – in order to

---

[21] David M. Lodge and Kristin Shrader-Frechette, "Nonindigenous Species: Ecological Explanation, Environmental Ethics, and Public Policy," *Conservation Biology* 17 (2003): 31–37. http://onlinelibrary.wiley.com.libproxy.sdsu.edu/doi/10.1046/j.1523-1739.2003.02366.x/pdf (accessed December 20, 2012).

[22] For a schematic outline of the kinds of scenarios in which assisted colonization might be beneficial and the studies needed to assess what can and should be done, see Luke P. Shoo, et al., "Making Decisions to Conserve Species under Climate Change," *Climatic Change* 119 (2013): 239–246.

[23] A comprehensive introduction to issues surrounding geo-engineering can be found in The Royal Society, *Geoengineering the Climate: Science, Government, and Uncertainty* (London: The Royal Society, 2009). http://royalsociety.org/policy/publications/2009/geoengineering-climate/ (accessed December 13, 2013).

cool the planet. The other involves removing CO_2 from the atmosphere, either technologically before it is released into the atmosphere or from the atmosphere by means, for example, of plant photosynthesis. Geo-engineering is neither a form of mitigation nor adaptation. It is not mitigation because it does not involve reducing emissions. It is not adaptation because it does not involve accepting the climatic consequences of greenhouse gas emissions in the atmosphere. Geo-engineering seeks to mitigate the climatic effects of greenhouse gas emissions without reducing the emissions.

Discussion of the merits of research into, and the use of, geo-engineering techniques have picked up recently as efforts to reach an international agreement to reduce greenhouse gas emissions have failed and emissions have continued to increase globally. Research into solar radiation management was given considerable credibility by the intervention of the Nobel Prize-winning chemist Paul J. Crutzen. Crutzen argues that, although mitigation is by far the best means for dealing with climate change, in absence of an international mitigation regime and in the presence of various environmental laws that reduce airborne particulate matter, which increase the Earth's albedo, "the usefulness of artificially enhancing earth's albedo and thereby cooling the climate by adding sunlight reflecting aerosol in the stratosphere might again be explored and debated."[24] The presence of aerosols, such as minute sulfur or hydrogen sulfide particles in the stratosphere, would reflect radiative energy into space and cool the planet. Volcanic eruptions, such as the 1991 Mt. Pinatubo eruption, confirm this cooling effect. Crutzen estimates that in order to compensate for the warming effect of a doubling of CO_2 in the atmosphere, we must inject about half the amount of sulfur into that stratosphere each year that the Pinatubo eruption blew into it. Injecting sulfur into the stratosphere is not the only form of solar radiation management under scientific discussion. Marine cloud brightening is also under discussion, but putting sulfur into the stratosphere is receiving the most attention currently.[25]

There is, of course, good reason to worry about the negative side effects of shooting sulfur into the stratosphere. Crutzen discusses the danger of ozone depletion. The sulfur produced by the Mt. Pinatubo eruption caused about a 2.5 percent loss of ozone.[26] The loss caused by injections of sulfur at high altitudes directly into the stratosphere, however, should be less. The warm

---

[24] Paul Crutzen, "Albedo Enhancement by Stratospheric Sulfur Injections: A Contribution to Resolve a Policy Dilemma," *Climatic Change* 77 (2006): 211–219. DOI: 10.1007/s10584-006-9101-y.

[25] For a recent discussion of marine cloud brightening, see John Latham, et al., "Marine Cloud Brightening," *Philosophical Transactions of the Royal Society* A 370 (2012): 4217–4262.

[26] Crutzen, "Albedo Enhancement," p. 215.

stratosphere could inhibit sulfur from forming into ice crystals, which are responsible for ozone holes. Moreover, because of the reduction of chlorofluorocarbon use, the ozone is healthier now than when Pinatubo blew its top. So, ozone reduction poses less of a threat than it once did.[27] Of course, we cannot be confident about the actual effects without a great deal more modeling and some physical testing.

Crutzen endorses a research-only policy, according to which research should proceed first by means of computer modeling and second through small-scale atmospheric tests.[28] Crutzen takes the usefulness of solar radiation management to be limited to cases in which "a low probability, high consequence" outcome starts to unfold. Presumably he is thinking of events such as the rapid melting of the Greenland ice sheet, which would increase sea levels by 7 m, or massive release of methane and CO_2 from the permafrost, which could constitute a catastrophic positive feedback in the warming process. In the words of energy scientist Jane C. S. Long, in the event of such a massive release, rather than the trickle that we are now experiencing, "it's game over."[29] As the planetary cooling that followed the Mt. Pinatubo explosion evinced, solar radiation management could produce results relatively quickly, within about six months. This increases its attraction as a tool in response to an emergency situation.[30]

Crutzen views the use of solar radiation management as only one among a suite of strategies if mitigation were insufficient. Increasing the Earth's albedo would cool the planet, but it would not stop the acidification of oceans caused by CO_2 uptake, which inhibits the calcification process necessary for the creation of shells on shellfish. In the absence of sufficient mitigation, we will also have to sequester CO_2 from the atmosphere and store it deep in the ocean or underground in a process known as *carbon capture and storage*.[31]

Although controversial in the scientific community, Crutzen's proposal for research into solar radiation management is supported by the President of the National Academy of Sciences, Ralph J. Cicerone. Cicerone also takes a research-only position. He advocates research that includes the identification of concepts, mathematical modeling, the anticipation of side effects, and publication in peer-reviewed journals.[32] He also urges a moratorium on

---

[27] *Ibid.*, 215–216.

[28] *Ibid.*, 215.

[29] Quoted in Michael Specter, "The Climate Fixers," *The New Yorker*, May 14 2012. http://www .newyorker.com/reporting/2012/05/14/120514fa_fact_specter (accessed December 23, 2012).

[30] Crutzen, "Albedo Enhancement," p. 216.

[31] *Ibid.*, 217.

[32] Ralph J. Cicerone, "Geoengineering: Encouraging Research and Overseeing Implementation," *Climatic Change* 77 (2006): 221–226.

what he calls "geoengineering interventions," by which he seems to mean large-scale use, but he makes provision for "field experiments."[33] As it happens, a moratorium on geo-engineering was established at the 2010 Conference of Parties of the Convention on Biological Diversity (to which the United States is not a party). Their decision seeks to

> Ensure, in line and consistent with decision IX/16 C, on ocean fertilization and biodiversity and climate change, and in accordance with the precautionary approach, that no climate-related geoengineering activities take place until there is an adequate scientific basis on which to justify such activities and appropriate consideration of the associated risks for the environment and biodiversity and associated social, economic and cultural impacts.[34]

However, the usage of "geoengineering activities" in this text is extremely vague. In addition to implementation, it could include research of any sort. The primary concern of the decision seems, in any case, to be directed toward ocean fertilization activities. It is unclear whether it should be interpreted to include solar radiation management.

The articles by Crutzen and Cicerone suggest an argument that I call *the research argument*, which I understand as follows: (1) Failure to mitigate climate change effectively could produce emergency situations in which it will be necessary to act fast to avoid climate catastrophes; (2) solar radiation management could possibly provide the only means for fast action; (3) the side effects of solar radiation management are unknown and possibly unacceptable; (4) only research into solar radiation management can provide us with knowledge about its side effects and how to control them; (5) we should pursue means for acting quickly and without terrible side effects to prevent climate catastrophes caused by our failure to mitigate, but not to the exclusion of other morally appropriate responses to climate change (e.g., mitigation and carbon capture and storage, both stressed by Crutzen); (6) therefore, we should pursue research into solar radiation management, but not to the exclusion of other morally appropriate responses to climate change.

The research argument might look unassailable, but moral philosophers, who have considered the matter carefully, raise several concerns about research into geo-engineering. To my knowledge, Dale Jamieson was the first moral philosopher to look carefully at the matter. While offering a lukewarm endorsement of research into solar radiation management, he states three

---

[33] *Ibid.*, 223 and 225.
[34] Conference of the Parties to the Convention on Biological Diversity, "Draft Decisions for the Tenth Meeting of the Conference of the Parties to Convention on Biological Diversity," 145. http://www.cbd.int/doc/meetings/cop/cop-10/official/cop-10-01-add2-rev1-en .pdf (accessed December 21, 2012).

concerns.[35] First, any money (and one might add time and research power) invested in geo-engineering takes resources away from other worthy research goals. So, the opportunity costs of research should be considered before deciding that it is all things considered justified. Second, an investment of resources into geo-engineering might have opportunity costs for other areas of climate change research in particular. Perhaps this is not the best approach to solving the problem of climate change. Third, research into a technology risks inadvertently using it. Thus, given both our current lack of understanding of the effects of the geo-engineering and the lack of any international agreement to sanction its use, the risks are significant.

Still, Jamieson does not take these concerns as establishing a compelling case against research into geo-engineering. I agree with that judgment. Although it is true that there are opportunity costs for any course of research, there are three reasons for thinking that these should not prevent research into geo-engineering. First, if a climate emergency arises that could have been remedied by the use of solar radiation management, but we are unable to offer the remedy because we have failed to do the research, the moral opportunity costs of not having conducted the research will be very high. Second, although we will need technological breakthroughs to eventually bring emissions to zero, the major hurdle to initiating an international agreement on mitigation is not a lack of adequate technology; therefore, research into solar radiation management will not take resources away from projects that might make an agreement possible. Third, generally it is too much to ask of a research project that it be justified in light of the foregone opportunities to pursue all the other possible projects. No research can be justified as being the best among all possible uses of research time and money. Unless there are specific opportunity costs that look high, simply invoking the existence of opportunity costs is not compelling.

Jamieson's worry that research will lead to too-early deployment is, of course, applicable to many areas of research. It's not clear why we should want to single out research into solar radiation management. If the reason is applied generally, there would be no progress in science and engineering. Thus, unless there is a reason to think that early deployment is especially likely in the case of solar radiation management, it would be unfair to block research into it on grounds that apply to all research. Moreover, leading scientists conducting

[35] See Dale Jamieson, "Ethics and Intentional Climate Change," *Climatic Change* 33 (1996): 333. Jamieson continues to support geo-engineering research. See his "Some Whats, Whys, and Worries of Geoengineering," *Climatic Change* 121 (2013): 527–537. DOI 10.1007/s10584-013-0862-9.

research into solar radiation management are reportedly not at all eager to see their research used.[36] This does not ensure that there will be no early use of solar radiation management, but there can be no such assurance with research into any form of new technology.

Stephen M. Gardiner expresses a different kind of moral concern about research into solar radiation management. The presence of moral corruption is the central theme of his detailed analysis of the theory and practice of climate change policy. His discussion of solar radiation management is just a small part of his important larger analysis.[37] He takes the corruption to exist in part because of three structural features of the problem of climate change. First, it is caused by billions of people diffused around the world and lacking in institutions to coordinate their activity. This creates a serious global collective action problem.[38] Second, the pernicious effects of climate change come mostly much later than their causes. This allows an earlier generation to emit greenhouse gases and to pass the buck along to later generations.[39] Finally, we lack the theoretical tools to think intelligently about long-term moral and practical issues such as climate change.[40] These factors encourage the globally affluent to "give undue priority to what happens within their own life time" and to "welcome ways to justify overconsumption and give less scrutiny than they ought to arguments that license it."[41] Such corruption is a cognitive and moral defect amounting to a kind of "self-deception."[42]

Gardiner is not directly concerned with whether research into solar radiation management or its use could ever be justified, but rather to unveil the presence of moral corruption in climate change policy, including policy that might be developed regarding the use of solar radiation management. If his account is plausible, it presumably makes it much harder to justify research into and the use of techniques such as solar radiation management. Gardiner assumes that solar radiation management is an evil, albeit perhaps a lesser evil

---

[36] Specter quotes Hugh Hunt of Cambridge University, saying, "I don't know how many times I have said this, but the last thing I would ever want is for the project I have been working on to be implemented . . . If we have to use these tools, it means something on this planet has gone seriously wrong." See Specter, "The Climate Fixers," *The New Yorker.*

[37] Stephen M. Gardiner, *A Perfect Moral Storm: The Ethical Tragedy of Climate Change* (Oxford: Oxford University Press, 2011), chp. 10.

[38] *Moral Storm*, pp. 24–32.

[39] *Ibid.* pp. 32–41.

[40] *Ibid.*, pp. 41–44.

[41] Stephen M. Gardiner, "Is 'Arming the Future' with Geoengineering Really the Lesser Evil? Some Doubts about the Ethics of Intentionally Manipulating the Climate System," in Stephen M. Gardiner, et al., eds., *Climate Ethics: Essential Readings* (Oxford: Oxford University Press, 2010), p. 287.

[42] *Ibid.*

than permitting insufficiently mitigated climate change to take its course.[43] The consequence of taking solar radiation management as an evil is that there is a special burden on those who would seek to justify using it. He contends that sometimes choosing the lesser evil can mar or even blight the life of the chooser. This is important in the context of research into solar radiation management because the point of such research would be to come to understand its effects and to evaluate its effectiveness for purposes of possibly making it available for future use. However, Gardiner's argument raises the disturbing thought that in doing so, we might be putting persons in the future in the position in which they would have to confront a choice about the employment of a lesser evil, and that such a choice could mar or blight them. It would seem better to avoid putting future persons in situations in which choices of this kind might be necessary.[44]

It should be clear that according to this approach, a great deal hangs on characterizing solar radiation management as an evil. Of course, nobody could credibly maintain that solar radiation management is simply the best approach for dealing with climate change. There are too many possible problems and too many unknowns. But that does not establish that it is an evil. It is useful, however, to distinguish between a lesser evil choice and a second-best choice. One difference is that being in the situation of having to choose the lesser evil is never desirable. But because the second-best choice is better than all of the others save the best, choosing the second-best might not be particularly unattractive. A second-best choice might even sometimes be easily made by ordering the alternatives according to how choice-worthy they are, and when the choice ranked highest is not available, proceeding down the list to the next.

The description of an easy second-best choice will, however, often be inaccurate. One reason for this is that the ordering of choices can be especially difficult. It can call for difficult judgments involving conflicting and incommensurable goods and rely on uncertain estimations of the probability of outcomes. For example, consider the following possibility: Substantial climate change mitigation, conforming to the antipoverty principle and allowing the right to sustainable development is the best choice. Moreover, after extensive research we believe with moderate confidence that the risks of solar radiation management are low. Now we must rank the second and third best from the remaining two options. One option is substantial mitigation that does not recognize the right to sustainable development. The other is less-substantial

---

[43] *Moral Storm*, p. 339.
[44] *Ibid.*, p. 389.

mitigation recognizing the right to sustainable development and a plan to use solar radiation management in the event of a developing catastrophe. Whether it is better to employ solar radiation management or not to recognize the right to sustainable development is not some facile choice. Whatever the second choice should be, it is far inferior to the first. Another reason a second-best choice may not be easy is that we might have especially strong reasons for seeking the best choice over the second-best choice, as the present example illustrates.

The relevant feature of an evil, as Gardiner understands it, is that we have good moral reasons to avoid it.[45] If we have a reason to avoid a policy because it involves doing evil, then advocates of the policy have the burden of showing why there are reasons for the policy that are sufficiently strong to defeat the reasons against it. Just war theory provides examples of this sort of reasoning. A war in pursuit of a morally important end is arguably sometimes the lesser evil in comparison to forsaking the just cause by not going to war. War is evil because it requires us to do things that are ordinarily gravely wrong, including intentionally killing other human beings and intentionally destroying some of the infrastructure that supports human well-being. Just war theory maintains that the presumption against going to war can be defeated if and only if the war satisfies certain stringent conditions. In contrast, it would seem to ignore the evils of war to refer to a particular war as a "second-best" means for achieving the just cause peacefully.

Because lesser evil policies are still evil, we should avoid circumstances in which we would be called upon to use them. St. Augustine, the first just-war thinker in the Christian tradition, refers to the world of war as "misery."[46] In the case of war, not only should we abide by strict necessary conditions for initiating a war, we should also seek to reduce the causes of belligerence in the world. We should regret the circumstances that require resorting to a lesser evil, even when it is the morally correct course of action. We should certainly avoid creating circumstances for others in which their best available choice is a lesser evil.

The relevance of all of this depends, however, on solar radiation management being an evil, albeit possibly a lesser one. On what grounds might we believe that we have good reasons to avoid using solar radiation management? What in the employment of solar radiation management plays the same moral

[45] *Ibid.*, p. 349.
[46] St. Augustine, *The City of God*, bk. XIX, chp. 7. In *Nicene and Post-Nicene Fathers*, First Series, Vol. 2. Translated by Marcus Dods. Edited by Philip Schaff. Revised and edited by Kevin Knight (Buffalo, NY: Christian Literature Publishing Co., 1887). http://www.newadvent.org/fathers/120119.htm (accessed July 28, 2012).

role that intentional killing plays in war – namely, the role of establishing a strong presumption against it? My sense is that Gardiner may be too readily accepting the dialectic of the existing debate, which assumes that solar radiation management is a lesser evil. He lists three considerations that incline people to believe that solar radiation management is an evil.[47] The first is that solar radiation management is at best a partial solution given that, for example, it does nothing to address the problem of ocean acidification. The second is that we are ignorant of its side effects. And the third is that it implicates humans further in the activity of dominating nature.

Regarding the first consideration, the fact that solar radiation management alone is an insufficient response to climate change is not a reason to avoid it. What of the second consideration? Epistemic uncertainty sometimes justifies precaution on minimax grounds. Application of the minimax rule to the question of whether to use solar radiation management would require us to compare the worst-case scenarios of solar radiation management to that of an unfolding climate catastrophe. But in order to take a possibility seriously as a worst-case scenario, there should be some causal antecedents in place such that we could understand its occurrence as the product of a causal chain following natural laws. It's not clear, when considering solar radiation management and a climate catastrophe, which possibilities are sufficiently connected with causal antecedents to take as real possibilities, but suppose as a catastrophe that the planet were to endure a 6°C increase in less than 100 years because of rapid methane release. That would be terribly bad. If the comparison is to making holes in the ozone, minimax might well direct us to favor solar radiation management.

The third consideration is concerned with the domination of nature. I addressed a similar concern when I considered the green objection to assisted migration. Our context is the anthropocene. We are not discussing a circumstance in which we either begin interfering with nature or we do not interfere at all. A warming induced by a massive release of methane would be an event caused by human activity. Hence, there does not seem to be a presumption against solar radiation management on grounds that it alone would constitute interference in nature. It is difficult to know if there is a sufficient common basis in values to argue with those who might say (as Gardiner reports some would say) that, "it would be better to all things considered to endure a climate catastrophe than to encourage yet more risky intervention."[48] In any case, endurance for a great many species, including ours, simply might not be an option.

---

[47] *Moral Storm*, pp. 348–349.
[48] *Ibid.*, p. 349.

If these three considerations are the reasons that many people consider solar radiation management to be an evil, then the conclusion that they reach does not seem warranted. The three considerations do seem, however, to give us reason to believe that the use of mitigation policies should be the first line of defense against climate change. Mitigation addresses the problem of ocean acidification. The uncertainties associated with mitigation are about whether it will be sufficient, not its side effects. And furthermore, mitigation seeks to lessen our impact on nature. It is beyond doubt that the best solution would be thorough mitigation. Perhaps solar radiation management is an evil, but we know so little currently about what its effects would be that we cannot confidently claim that. Given the uncertainty about warming at any level of temperature increase, the most reasonable approach would seem to include robust mitigation as well as research into solar radiation management, and perhaps even into other forms of geo-engineering.

If we had the technical capacity to employ geo-engineering within morally acceptable risks, its legitimate use would require an appropriate regulatory framework and decision-making process. Some sort of international agreement sanctioning its use only under a set of carefully established conditions would be required.[49] It would be at best a lesser evil to use solar radiation management in an emergency situation without satisfying the requirements of legitimacy, but we need not suppose that every use of solar radiation management would be a lesser evil. Much greater attention needs to be paid to the regulation of the use of geo-engineering. The United Kingdom has endorsed a set of principles, called *The Oxford Principles*, but these are insufficiently specific for regulatory purposes.[50] Recognition of the problem of the legitimate use of solar radiation management provides a reason to begin diplomatic discussions about the matter, not a reason to believe that geo-engineering is an evil.

I can imagine that someone in the grips of concern about moral corruption might doubt some or all of the responses I have offered on grounds that philosophical reflection on solar radiation management is beset by moral corruption and is subject to self-deception. The problem with that suggestion, however, is that if we doubt positions not as the result of raising specific objections to the truth of the claims or the logic of the argument, but on the general suspicion of self-deception, there is no way to distinguish what is legitimately doubtable from what is not. Everything is called into question and we have no rational means for settling anything. Not surprisingly, affirmation

---

[49] See also Daniel Bodansky, "The Who What, and Wherefore of Geoengineering Governance," *Climatic Change* 121 (2013): 539–551. DOI 10.1007/s10584-013-0759-7.

[50] See The Oxford Principles. http://www.geoengineering.ox.ac.uk/oxford-principles/principles/ (accessed December 13, 2013). For a discussion of the Oxford Principles see Steve Rayner, et al. "The Oxford Principles," *Climatic Change* 121 (2013): 499–512. DOI 10.1007/s10584-012-0675-2.

of the doctrine of total depravity has not been the basis of a lot of philosophical inquiry.

If solar radiation management were not a lesser evil but instead a second- or third-best policy option, it would still require moral justification. The justification would be to show that the better and best policies were not available. If there is nothing inherently evil about solar radiation management, then absent considerations of moral hazard, there would be nothing regrettable about research into geo-engineering to consider its effectiveness and possible side effects in the event that better options are not be available. On the contrary, preparing a Plan B often is the responsible approach, especially if there are good reasons to worry that Plan A will be unsuccessful.

Whether or not there would be an evil, albeit possibly a lesser one, in solar radiation management would seem to depend mostly on the risks associated with attempting it and on the procedures used when deciding to employ it. We can know the former only by engaging in scientific research, including possibly small-scale tests. The only hope for having adequate decision procedures in place is to begin the diplomatic process that would lead to institutions that would regulate the use of solar radiation management technology. These considerations simply underscore the research argument implicit in articles by Crutzen and Cicerone.

In one way, however, the research argument might be too optimistic. Recall that the first premise is the following: (1) Failure to mitigate climate change effectively could produce emergency situations in which it will be necessary to act fast to avoid climate catastrophes. In fact, there are cascading uncertainties attached even to low temperature increases, and the magnitude of the possible catastrophes is enormous. Moreover, there is already evidence of some of the causal antecedents of these catastrophes, such as the methane percolating from arctic waters. Because methane is thirty times stronger than CO_2 as a greenhouse gas, a massive release of methane would be catastrophic. Contrary to the subjunctive mood of (1), we might have already failed to mitigate climate change effectively at the current temperature. There seem to be very good precautionary reasons to pursue the research into solar radiation management and to begin the diplomatic efforts to regulate its deployment.[51]

A MORALLY SATISFACTORY INTERNATIONAL MITIGATION AGREEMENT

The discussion of adaptation, assisted migration, and geo-engineering has shown that although each of these may be an important supplement to an

---

[51] The view that geo-engineering can complement mitigation efforts is also defended by Andy Ridgwell, et al., "Preface: Geoengineering: Taking Control of Our Planet's Climate?" *Philosophical Transactions of the Royal Society A* 370 (2012): 4164.

effective international mitigation plan, none is a substitute. Given the morally constrained emissions budget, the urgent task of developing an effective international mitigation framework remains. Any morally satisfactory international mitigation agreement should satisfy at least the following four conditions:

(1) There should be a rational relation between the goal of mitigation and the means prescribed internationally. This includes primarily a schedule of emissions reductions reasonably likely drastically to reduce global emissions within a period of three to four decades.

(2) The plan should encourage cutting subsidies for the use of coal, putting a price on CO_2 emissions, and subsidizing the use of renewable energy. It should, in other words, sharply curtail greenhouse gas emission by promoting a technological revolution in energy, not by a steep reduction in economic activity.

(3) The international framework for energy use should be fair. This includes safeguarding the right to sustainable development by offering credible assurance that the price of energy in all of the least developed and many of the developing countries will not increase due to requirements of the mitigation plan.

(4) The process of developing the plan and the plan itself should be adequately responsive to the urgency of the need to slow the consumption of the morally constrained carbon budget.

The first condition is simply a minimal requirement of rational action – namely, that actors should adopt the means that are sufficient for the end intended. When the intended end is morally required, the first condition does more than state a rational constraint on action. It also states a moral constraint. The second condition concerns the possibility of achieving emissions reductions by either reducing global economic activity or developing substitutes that involve fewer or no emissions.[52] The moral problem with the former approach is that it is recessionary. A global economic recession would produce terrible suffering. As the global economic slowdown starting in 2008 demonstrated, people living in developing countries are strongly affected by recessions in the developed world because their economies are reliant on investment and lending originating in the developed world. People living in least-developed countries are affected because their economies are often fueled by basic commodity exports to, and remittances from, more developed countries. Many

[52] See also Henry Shue, "Avoidable Necessity: Global Warming, International Fairness and Alternative Energy," in Ian Shapiro and Judith Wagner DeCew, eds., *NOMOS XXXVII: Theory and Practice* (New York: NYU Press, 1995), pp. 247–249 and 257–259.

people in developing and least-developed countries are especially vulnerable in economic downturns because of their already low incomes and the lack of state capacity to provide social welfare benefits. Hence, any morally acceptable climate change mitigation treaty must be devoted to facilitating a technological transformation in energy, not a reduction of global economic activity.

The third condition draws on the arguments developed in Chapters 5 and 6. Safeguarding the right to sustainable development is an important moral constraint on a climate change treaty and to ensure that it is protected, states should be required to mitigate on the basis of their ability to pay for global climate change mitigation. The fourth condition is a reminder that the clock is ticking. At the current level of global emissions, by 2040 the morally constrained emissions budget for limiting warming to 2°C will be expended. In the meantime, massive investments are being made – approaching $700 billion in 2012 – to find and extract fossil fuels from under the ground.[53] The politically powerful fossil fuel lobby intends to delay establishing a price on CO_2 emissions.

How realistic is it to suppose that these constraints can be satisfied? Broadly speaking, there are three scenarios that might result from climate change negotiations. (1) An international climate regime that is sufficiently demanding of global reductions to be reasonably likely to limit warming to 2°C, but that raises energy prices in the least-developed and developing countries so that the third constraint is violated. (2) A regime that is sufficiently demanding of global reductions to be reasonably likely to satisfy the 2°C limit that does not significantly raise energy prices in the developed and least-developed countries, thereby satisfying all three constraints. This is the regime that is our best hope for avoiding dangerous climate change. Or (3) a regime that does not significantly raise energy prices in the developed and least-developed countries, but is insufficiently demanding of global reductions to be reasonably likely to meet the 2°C warming limit, thereby violating the first constraint. Currently, negotiations seem to be drifting in this direction.

The richest countries have two strategic bargaining advantages that could make an agreement such as the first kind seem most likely. These include their geographic location and their greater capacity to fund adaptation. The greatest human suffering caused by climate change is projected to occur in poor countries, not the wealthiest.[54] Moreover, the richest countries have

[53] See Carbon Tracker, "Unburnable Carbon."
[54] See United Nations Human Development Programme, *Human Development Programme 2007/2008 Fighting Climate Change: Human Solidarity in a Divided World*, pp. 24–31.

much greater means to implement infrastructural plans for adapting to climate change. Wealthy countries might then decide that it is in their interest to forgo a strong international mitigation agreement and spend on adaptation instead – or at least to pose the credible threat of doing so. Such threats could make weaker and poorer countries more willing to accept proposals of the first type. Such reasoning might explain the United States' resistance to joining Kyoto, its aversion to binding emissions limits at COP 13 in Bali in 2007, and its failure to lead at COP 15 in Copenhagen in 2009.

In contrast to this sober analysis, some observers see a reason for hope in the growing influence of some developing countries, especially those organized in the BRICS group, comprising Brazil, Russia, India, China, and South Africa. As long as the countries of the developed world see the value of reaching an effective mitigation agreement, they will be eager to draw in the countries of the developing world, especially those with large and quickly growing economies emitting a great deal of CO_2. This gives at least some developing countries more bargaining power than they would have in international negotiations over policies in which their participation is less needed. Because China is now the single largest total emitter, responsible for nearly 20 percent of the annual global emissions, the Annex-1 countries need its participation for there to be any meaningful path toward a global emissions reduction. Some observers take the bargaining power of developing countries to be a reason for optimism that an agreement on a proposal of the second type could be achieved.[55] According to the optimists, climate change fundamentally alters global power relations, making realistic the possibility of more-just global institutions generally.

The hopeful conclusion is not, however, the only one that could be drawn from the previous analysis. One reason for tempering optimism derives from a structural feature of the international negotiations, which by metaphysical necessity renders members of distant future generations unable to veto proposals of the third type or to prevent the global community from drifting in this direction in the absence of any agreed upon proposal. But representatives of states that would be made to carry significant burdens are at the table to veto

http://hdr.undp.org/en/media/HDR_20072008_EN_Complete.pdf (accessed April 29, 2013); Intergovernmental Panel on Climate Change, "IPCC, 2007: Summary for Policymakers," in M. L. Parry, et al., eds., *Climate Change 2007: Impacts, Adaptation and Vulnerability. Contribution of Working Group II to the Fourth Assessment Report of the Intergovernmental Panel on Climate Change* (Cambridge: Cambridge University Press, 2007), pp. 11–18, http://www.ipcc .ch/pdf/assessment-report/ar4/wg2/ar4-wg2-spm.pdf (accessed April 29, 2013); and the World Health Organization's submission to the Ad Hoc Working Group on Long-Term Cooperative Action, http://www.unfccc.int/resource/docs/2009/smsn/igo/047.pdf (accessed April 29, 2013).
[55] See Tom Athanasiou and Paul Baer, *Dead Heat: Global Justice and Global Warming* (New York: Seven Stories Press, 2002).

proposals of the first and second types.[56] Given the magnitude of the global emissions reductions necessary to limit warming to 2°C, there is a tendency to drift toward the third scenario.

Additionally, the lesson of the personal analogy discussed in Chapter 1 is that whether a risky course of action is properly judged as dangerous depends on the moral opportunity costs of avoiding it. If these costs are sufficiently high, there are more compelling moral reasons to pursue the risky course of action than to avoid it. Hence, developing and least-developed states have reason to look on climate change as dangerous only if the moral opportunity costs to them of mitigation are less than those of adaptation. If an international mitigation plan would slow human development and prolong the desperate poverty that billions of people live in, it is not obvious that the representatives of poorer states should consider the plan as the least-dangerous alternative. This gives these representatives a reason to reject proposals of the first type, which prolong their energy poverty. Because decisions within the UNFCCC require unanimity, this is further reason to believe that the likelihood of a proposal of the first type becoming the basis of a new international regime is very low. Either agreement on a proposal of the second type – the best-case scenario – or drifting in the direction of the third possibility seems likely. The gravest danger is in drifting.

The four conditions previously discussed apply directly to an international climate change agreement. More indirect are the conditions of a morally satisfactory domestic policy that might be developed in light of the conditions for an international agreement. These include the following:

(5) Domestic plans to contribute to climate change mitigation should be made in light of the state's duty to respect the right to development and to fulfill its responsibility in light of its ability to pay.

(6) The domestic implementation of the plan in countries in which coal mining and burning are extensive should include retraining for workers in these industries and decent alternative job opportunities.

(7) The domestic implementation of the plan in highly developed states in which the price of energy may temporarily rise should include subsidies for the provision of heating and cooling for the poor.

Although the fifth condition concerns the development of a domestic mitigation policy, it is related to states carrying their fair share in light of the

---

[56] This is an instance of the intergenerational problem analyzed in great detail by Stephen M. Gardiner in his *A Perfect Moral Storm: The Ethical Tragedy of Climate Change* (Oxford: Oxford University Press, 2011), pp. 143–184.

third condition to develop a fair mitigation plan. But it seems unlikely that a domestic plan will satisfy the fifth constraint if the policy is responsive only to the demands of the citizenry of the state, and not to bargaining pressures from other states. One way to understand this problem is in terms of what might be called *a community of justification*, which is the group of persons to whom a course of action or policy must be justified. One important feature of the public justification of policies and political principles is that it exposes them to the interpretations of values, the experiences, and the interests of the public and thereby helps to overcome various limitations in the process of justification, including narrowness of experience, lack of adequate foresight, and discounting the interests of others.[57] If the community of justification does not include everyone with an important moral stake in the policy, then the likelihood of the process of justification adequately transcending the narrowness of its community is diminished. If, for example, U.S. climate policy is justified only to U.S. citizens and not subject to the bargaining pressures of other states, then it is far less likely that it will be adequately responsive to the responsibilities the United States should shoulder in an international mitigation agreement. One morally important feature of international bargaining is the broadening of the community of justification. Although critics might decry the formation of policy in light of international bargaining pressure as undemocratic, under the system of sovereign states, this is an important means by which the community of justification can be broadened.

The sixth and seventh conditions derive from the concerns of the antipoverty principle applied to the domestic effects of climate change policy. The poor should not be made to pay the costs of the transition to an economy based on renewable energy if there are alternatives. These requirements suggest the need for states to collect and redistribute the revenues that flow from putting a price on CO_2 emissions. Both a tax on emissions and a system of capping and trading emissions entitlements are superior to command-style regulation in this regard. Taxing emissions or selling initial emissions entitlements creates public funds that can be used to offset the domestic costs of the transition to an economy based on renewable energy. In recent political debates in the United States, critics of the cap and trade plan characterized it as "tax and trade" and managed to scare a large number of Americans into believing that it would erode their incomes. This undermined support for the policy in the U.S. Senate. But the analysis of the U.S. Congressional Budget Office reveals that the net costs of such a policy could be beneficial to

---

[57] See also Elizabeth Anderson, *Value in Ethics and Economics* (Cambridge, MA: Harvard University Press, 1993), p. 111.

low-income Americans, as long as initial entitlements are sold and the revenue generated is redistributed appropriately.[58] Satisfaction of conditions six and seven, then, is well within the reach of policy makers. And in order to build broad popular support for climate change policy, satisfying these constraints is necessary.

PLEDGE AND REVIEW

The failure to achieve an effective climate change mitigation plan by means of an international negotiation process in which states centrally negotiate their share of an overall emissions reduction burden has lead some states, notably the United States, and some students of policy to advocate alternatives that are more decentralized, such as pledging reductions derived independently of an international negotiation process and subjecting these pledges to periodic reviews. Given the negotiation failures to date, it is reasonable to seriously consider alternative approaches. The moral guidelines for considering them, I suggest, are the four requirements of a morally satisfactory international mitigation agreement discussed in the previous section.

The Copenhagen Accord formulated at COP 15 in 2009 includes a list of pledges by countries to reduce their CO_2 emissions by 2020. These are voluntary pledges that are entirely the product of domestic political debates. In light of the importance of broadening the community of justification, it is hardly surprising that the total reductions pledged by this means is far below that which is likely to be required to stay within the 2°C warming limit. The United Nations Environmental Program's (UNEP) study of the consequences of the emissions reductions pledged in Copenhagen found that the temperature increase they would be likely to produce would be in the range of 2.5°C–5°C.[59] As it currently stands, the Copenhagen approach fails the first requirement: that the means be sufficient to the end.

Another serious problem with the Copenhagen approach is recognized by defenders of more robust versions of pledge and review. David G. Victor emphasizes the need for thorough, transparent reviews, with consequences

---

[58] Congressional Budget Office, "The Estimated Costs to Households from the Cap-and-Trade Provisions of H.R. 2454." June 19, 2009. http://democrats.energycommerce.house.gov/sites/default/files/documents/CBO-Household-Effects-Cap-and-Trade-2009-6-19.pdf (accessed December 10, 2013).

[59] United Nations Environmental Program, *The Emissions Gap Report*, 2010. http://www.unep.org/publications/ebooks/emissionsgapreport/pdfs/GAP_REPORT_SUNDAY_SINGLES_LOWRES.pdf (accessed November 22, 2012).

for failures to meet targets.[60] Because of the costs associated with affecting a thorough transition to an economy based on renewable energy sources, states are unlikely to commit to the process without assurances that other states will do so as well. A process of pledge and review in which there are no sanctions for states which, upon review, are failing to meet their pledges will not provide sufficient assurance to states that their economies will not suffer competitive setback if they take the risk of assuming the costs of emissions reductions. There are a wide range of consequences that might be attached to failure to achieve pledges, including loss of various benefits that derive from being a member in good standing of a mitigation agreement, loss of forms of credit used to reward states that make reductions, and various sanctions on trade. The basic idea is that there should be some incentives to joining a pledge and review agreement and that some of these benefits will be conditional on performance.

Versions of pledge and review that are responsive to the need to broaden the community of justification and that employ meaningful reviews with consequences deserve serious attention. However, one serious problem with ideas of this sort is the time that it might take to navigate the complicated negotiations process of developing various forms of credit and sanction, attracting states to join in the process of pledging, and getting policies to converge around sharp emissions reductions. Victor draws analogies to the development of the World Trade Organization, which he acknowledges has lost all momentum in the Doha round.[61] It is far from clear that development of an effective pledge and review process can develop fast enough to satisfy the moral urgency expressed by the fourth constraint.

The problem of urgency is pervasive in climate change policy. The Durban Platform for Enhanced Action, agreed upon at COP 17 in Durban, includes a commitment "to develop a protocol, another legal instrument or an agreed outcome with legal force under the Convention applicable to all Parties."[62] Although this offers the hope of some sort of binding international instrument with more ambition than the pledges in Copenhagen, even if the process stays on schedule, the outcome will come into force very late. The Durban Platform

---

[60] David G. Victor, *Global Warming Gridlock: Creating More Effective Strategies for Protecting the Planet* (Cambridge: Cambridge University Press, 2011), pp. 252 and 257–259.

[61] *Ibid.*, p. 260.

[62] United Nations Framework Convention on Climate Change, *Report of the Conference of the Parties on Its Seventeenth Session, Held in Durban from 28 November to 11 December 2011.* http://unfccc.int/resource/docs/2011/cop17/eng/09a01.pdf#page=2 (accessed November 22, 2012).

calls for the plan to be completed no later than 2015 and to enter into effect only in 2020.

HOPE

If the world waits until 2020 to begin reducing total emissions, steep reductions will be required in order to extend the length of the budget significantly beyond 2041. Although this should have the effect of encouraging a technological transition, which would continue the work of extending the budget, such a massive and quick transition will be very expensive. The forces with the strongest interests against change will be well financed and will advocate against assuming such costs. Resisting them will take tremendous political organizing skills on the part of civil society movements and political organizations. This makes the current efforts to mobilize support for climate change mitigation in states currently lacking in mitigation ambition enormously important, so as both to begin mitigation sooner rather than later and to build a base of support to resist the fossil fuel industry and its political allies in the coming years. International campaigns such as the one organized around the Declaration on Climate Justice (see Appendix D) also have important roles to play in drawing attention to the need for a fair and more ambitious mitigation regime. The gravest dangers to that project are doing too little and doing it too late. Against these dangers, the efforts to mobilize citizens on behalf of mitigation sooner rather than later are our best hope.

Afterword: Frankenstorms

When someone at the National Oceanic and Atmospheric Association referred to the confluence of Hurricane Sandy with a polar trough just before Halloween as a "Frankenstorm," it may have been largely owing to the storm's intensity and timing. But "Frankenstorm" produces another association: This monster, like the one in Mary Shelley's novel, may have been human created.

The Intergovernmental Panel on Climate Change (IPCC), summarizing data gained from countless observations, has reported that over the hundred-year period from 1906 to 2005, the average global surface temperature increased 0.74°C. Although the oceans have warmed less quickly than land areas, they have been taking in more than 80 percent of the heat being added to the climate system. Warm ocean waters are the energy supply for cyclones. This is why such storms lose force when they run aground. And according to the IPCC, "[t]here is observational evidence of an increase in intense tropical cyclone activity in the North Atlantic since about 1970."

Has the warming of the oceans caused more intense tropical storms? Are greenhouse gas emissions the creator of a Frankenstorm? Oceanic warming is consistent with intense tropical storms, but it is not possible to run the kind of test needed to infer causation in any particular case with a sufficient degree of certainty. John Stuart Mill described a "method of difference" to determine if a certain factor caused an event; remove the factor while keeping the other factors in place and see if that leads to an alternative outcome. But we cannot rerun history over the last 150 years and take CO_2 emissions out to make the comparison with today. And even if we could, many other things would be very different without a coal-powered industrial revolution.

Intense tropical cyclones would presumably have formed in the absence of the current warming of the oceans by the greenhouse effect. Maybe Sandy would have been one of those. But the fact is that Sandy is not one of those, but one that formed after the ocean had warmed because of human activity.

And Sandy follows a summer of record heat waves and droughts in the United States and polar ice retreating to its lowest level ever recorded.

It is worth reminding ourselves of Victor Frankenstein's response to the terror that he unleashed. His life was ruined by regret – and why not? The monster that he created was directly or indirectly responsible for multiple deaths. Of course, these deaths are bad no matter how they came about, but to see one's own hand in them is horrific. Frankenstein was driven to do something about the monster – if not to protect others, then at least to have revenge.

The possibility that the damage Sandy has caused, and that any future tropical cyclone causes, could be the result of human doings should be an occasion for regret on the part of our political leaders; regret for their failure to bring about an international climate change treaty that makes any contribution to stabilizing the concentration of greenhouse gases in the atmosphere. The mere possibility that the damage this storm has wrought could be caused by our emissions should be enough to spur our political leaders to act.

Even if this one is not a real Frankenstorm, there is little doubt that such storms are forecasted. Were we to stabilize greenhouse gases in the atmosphere today, the thermal inertia of oceans would result in warming for centuries to come. The monster is on the loose. It is a mark of Victor Frankenstein's humanity that he was profoundly disturbed after the suffering caused by his monster. Will we see any such humanity in our political leaders once Sandy is gone?

The Antipoverty Principle and the Non-Identity Problem

Derek Parfit sees a problem for moral theories comprising principles he calls "The Person-Affecting View," or simply V, which holds that "it is bad if people are affected for the worse."[1] He is using "worse" here as he uses "worse-off," which he takes to refer "either to someone's level of happiness, or more narrowly to his standard of living, or more broadly, to the quality of his life."[2] To claim that a person is affected for the worse requires a comparison between the person's well-being as she is affected by the action and as she might have been without the action.

Changes in some institutional frameworks affect who will live in the future. As we adjust our energy policy, we will affect an array of major components of modern society from transportation to industrial production to recreation. These affect the circumstances in which people meet, have romantic encounters, and conceive children. The non-identity problem arises for theories such as V in these cases because they cannot account for what is wrong (when there is a wrong) in cases where policies affect who will be brought into existence. These policies do not render actual future persons worse-off because without such policies these persons would not have existed.[3] We can call that the argument *the no worse-off argument*.[4] The non-identity problem is sometimes misunderstood as a kind of skepticism about whether there can be anything wrong with actions in which the identity of persons is contingent on what we do. It is not that. Rather, it is a claim about the limitations of certain moral theories in accounting for what is wrong. It is, in other words, a problem for normative ethics, not applied ethics.

---

[1] Derek Parfit, *Reasons and Persons* (Oxford: Oxford University Press, 1984), p. 370.
[2] *Ibid.*, pp. 357–358.
[3] *Ibid.*, pp. 361–364.
[4] Following James Woodward, "The Non-Identity Problem," *Ethics* 96 (1986): 130.

In Chapter 1 I defended the following antipoverty principle as part of an identificatory theory of dangerous climate change:

> Policies and institutions should not impose any costs of climate change or climate change policy (such as mitigation and adaptation) on the global poor, of the present or future generations, when those costs make the prospects for poverty eradication worse than they would be absent them, if there are alternative policies that would prevent the poor from assuming those costs.

In this appendix I consider whether the antipoverty principle, in its rejection of policies that lay costs on the future poor, is undermined by the reasons that Parfit offers to reject V. Is the antipoverty principle a version of V? If so, then apparently it is undermined by the non-identity problem.

Now, V holds that it is bad if people are affected for the worse. One way to read V is as making a claim about states of affairs. It states the conditions in which states of affairs are bad: all other things being equal between two states of affairs, one contains more badness than another if people's well-being is worse in the one than the other. In contrast, the antipoverty principle judges policies and institutions, and is therefore not person-affecting in this sense of V. But it is more charitable to read V as applicable to principles concerned also with actions, policies, and institutions. This can be done by supposing that Parfit is assuming a piece of teleological moral theory, namely that it is wrong to do what is bad. More simply, we can read "it is bad" as equivalent to "an action is wrong." So read, "V says that an action is wrong if the state of affairs that it produces makes persons worse-off than an alternative action would." Read this way, V fits nicely into a consequentialist normative ethical theory.

On a standard interpretation of consequentialism, the antipoverty principle is non-consequentialist because it entails that an action can be right even if it is not one that would produce the optimal state of affairs. This can be seen if we set aside the non-identity problem and (unrealistically) assume stable identities of person across different possibilities. According to the antipoverty principle, policies and institutions that regulate the emissions of CO_2 are to be avoided if they lay avoidable costs on the poor, costs that prolong their poverty, even if those policies and institutions produce the optimal distribution of well-being. Hence, the antipoverty principle does not hold that policies and institutions should be avoided just because some persons are rendered worse off. It is not person-affecting in that sense.

Still it is not obvious that V includes only consequentialist theories for even non-consequentialist theories can require comparisons between the states of persons under two different actions. And the non-identity problem has purchase on V because it requires comparing the conditions of the same persons in two different circumstances. But the antipoverty does not seem to possess the features that make V problematic in non-identity cases. Insofar as it is a principle for the evaluation of actions or institutions, it does not require a comparison between the same persons' conditions as they are and as they might have been. The antipoverty principle is not subject to the problems of V.

In response to this defense of the antipoverty principle, it might be objected that the problem with person-affecting views, namely that they require a comparison where none can be obtained, afflicts any meaningful principle that evaluates actions, policies, or institutions by their impact on individuals, either with respect to aggregate well-being or with respect to the prospects for poverty eradication.[5] Although the antipoverty principle appears to avoid problematic comparisons of the same persons, it either rests on them or is nonsensical. Consider the following argument:

1. To claim that persons are wronged requires, at least implicitly, a comparison to a possible state of affairs in which these same persons would not have been wronged, regardless of whether the wrong is understood as a comparative diminution of the person's well-being or as a failure to act in accordance with non-consequentialist standards.
2. Such comparisons are undermined by the non-identity problem.
3. The antipoverty principle states a sufficient condition for claiming that persons are wronged by climate change related institutions.
4. Therefore, the antipoverty principle is undermined by the non-identity problem.

The first premise of this argument in effect broadens the class of person-affecting views to include non-consequentialist principles. It is making a conceptual claim about what "wronged" means. If "wronged" is coherently employed, it must be the case that there is an alternative action or practice in which the identical persons would not be wronged. If that were true, premise one would either reject the antipoverty principle as incoherent for not making the requisite comparison or bring it into the fold of problematic person-affecting views.

Premise one is, however, implausible. There are well-known and much contested accounts of wrongdoing or injustice that do not satisfy the requirement it states. Although these accounts might be false, the debate that surrounds

---

5   Edward Page seems to take this view in *Climate Change, Justice and Future Generations* (Cheltennam: Edward Elgar Publishing, 2006), pp. 138–150.

them is normative, not conceptual. It is implausible to think that they are false simply because they are incoherent. Consider, for example, John Rawls's difference principle. This principle requires that institutions maximize the expectations of the least advantaged members of society. Employing the Difference Principle necessarily involves comparing institutional arrangements to assess in which the least advantaged would be best-off. But the principle does not assume that the least advantaged will be the identical persons across the comparisons. This is obvious when one reflects on how Rawls identifies the least advantaged. He takes three characteristics as salient: those who are most disadvantaged in terms of family fortune and class origin; those whose natural endowments permit them to fare less well; and those whose luck in life results in less happiness.[6] It is significant that each of these three characteristics is basic-structure-dependent. In other words, whether a person is characterized in any one of these ways depends on how the social institutions of society assign benefits and burdens. Under different institutions, different people would be least advantaged. So, if we were to compare an actual society with a possible alternative that possessed very different institutions – for example, the actual one being libertarian and the possible alternative being egalitarian – different persons would be least advantaged. The Difference Principle directs us to compare the well-being of these different people, not identical people in different arrangements. Now, of course, the Difference Principle might be an invalid principle of justice. Certainly it is controversial. But the controversy that it has generated is normative; it concerns whether the principle's requirements are appropriate. It is hard to understand how the Difference Principle could generate such controversy if it were simply incoherent.

If premise one is implausible, then a moral principle that applies to future generations need not compare different conditions of identical persons in order to sensibly assert that persons have been wronged by institutions regulating CO_2 emissions. And the fact that the antipoverty principle does not do that does not make it incoherent. If there is plausible criticism of antipoverty principle, it is normative, not conceptual.

Parfit, however, has two arguments against invoking non-consequentialist principles to avoid the non-identity problem. In one he criticizes invoking rights to solve the problem. But the antipoverty principle does not invoke rights. So, I wait to discuss that criticism in Appendix B. He also contends that plausible non-consequentialist accounts of wrongdoing either must take the wrong-making feature of an action to consist in a failure of the action

---

[6] John Rawls, A *Theory of Justice*, rev. ed. (Cambridge, MA: Harvard University Press, 1999), p. 83.

to gain consent from those affected by the harmful action or require that those affected regret the action. But in cases of institutional effects on future persons, consent cannot possibly be obtained. So, the institutions cannot be criticized for failing to be the objects of consent. And it is also not the case that future persons whose lives are worth living, but who suffer from poverty that might have been avoided by a different energy policy, would necessarily regret our institutions. The reason for this is that they would not have existed but for these institutions.[7] So, if there is no failure of consent (because no consent is possible) and no regret, then according to Parfit's understanding of non-consequentialism, there can be no wrongdoing. Non-consequentialist accounts, then, cannot be used to judge policies in cases in which the identity of persons is contingent on what we do.

Parfit does not analyze the concept of regret in any detail. So it is somewhat difficult to assess the criticism that because of the absence of regret (where consent is not possible), there is no non-consequentialist basis for criticizing the action or policy. I start by noting that regret is an attitude, which either necessarily involves a moral judgment or not. Either way, however, it seems doubtful that persons who owe their existence to institutions could not regret them.

One possibility is that to regret a past action or practice necessarily requires both the judgment that it was wrong and the belief that the course of history would have been morally better without it. Consider the case of American slavery. Many – perhaps most – present citizens of the United States whose ancestors were in the country during the time of slavery would not have existed but for the institutions of slavery. This is plausible with respect to descendants of slaves, but, when we consider the scale of related events such as the Civil War, it is also plausible with respect to a very many of those whose ancestors were not slaves. My father's grandfather was a German immigrant who fought on the Union side of the Civil War. Presumably his life would have been very different had there been no slavery. Would he have met and married the same woman? Would they have conceived of my grandfather? The overwhelming majority of the people now alive whose ancestors were in the country at the time of slavery judge slavery to have been a gross injustice, even though a great many of them would not exist absent the slavery. This is analogous to persons who judge that their share of climate-change-related costs is unjust. The fact that I quite probably would not exist but for slavery does not make it incoherent for me to judge slavery to have been a grave injustice, and it does not prevent me from regretting slavery in that sense. Such moral judgments are not incoherent. In affirming the worth of our lives, we need not affirm

---

7 Parfit, *Reasons and Persons*, pp. 364–366.

everything that was necessary to bring it about. Just as when we admire the Great Wall of China, we may still ask, along with Bertolt Brecht, "In the evening... where did the masons go?"[8]

Perhaps the example of people whose existence is contingent on slavery does not capture what is deficient with non-consequentialist accounts in non-identity circumstances because in the example the descendants are recognizing an historical injustice but not an injustice to them. Could it be that they can rationally regret the injustice, even though they would not have existed absent the injustice, only because the injustice is not to them? Perhaps the problem arises only in the case of a person claiming not merely that a wrong has existed, but that *she* has been wronged, even when she would not exist but for the allegedly wrong act or practice. This, however, still seems false for more or less the same reasons. There does not seem to be anything obviously incoherent about a person judging that, in some respects, the world would have been a morally better place if the practice to which, in part at least, she owes her existence had not existed.

There might be another reason that regret about a practice is unavailable to future persons whose existence is contingent on the practice. Perhaps it is the case that a self-respecting person cannot wish that a practice to which he owes his existence not have occurred. But if that is the problem regret poses to non-consequentialist views in non-identity cases, it is not unique to such views. Presumably, for all moral theories it should be the case that that which is judged wrong can be regretted. But if self-respect does not permit condemning a practice to which one owes one's life, then there is also a problem for Parfit's partial answer to the non-identity problem that applies only when the outcomes compared contain the same number of people. Imagine that in the future self-respecting persons are less well-off than the same number of other persons might have been if different institutions had regulated CO_2 emissions. Although these persons have been taught *Reasons and Persons* in school, and are therefore able to identify the historical wrongness of regulatory institutions, according to the present criticism they also cannot sincerely regret the injustice insofar as they are self-respecting. If there is a way out for Parfit's principle, it is by denying that a self-respecting person cannot sincerely wish the nonexistence of institutions on which her existence depends. But, of course, this way out could be used by the antipoverty principle as well.

Maybe the point about regret is not that persons who would not exist but for a putatively unjust practice cannot experience regret, but that they might not,

---

[8] Bertolt Brecht, "A Worker Reads History." Available online at: http://allpoetry.com/poem/8503711-A_Worker_Reads_History-by-Bertolt_Brecht (accessed December 26, 2012).

and when they do not, there is no injustice on a non-consequentialist account. But it is difficult to see why the attitude of the putatively wronged person should be a necessary condition of there being an injustice. This is not a standard feature of non-consequentialist accounts. According to Rawls, for example, the least advantaged do not need to regret the social institutions that have governed their fate in order for the institutions to be unjust. Part of the point of contractualist accounts, which take justification to be a function of some sort of idealized or hypothetical consent, is that persons' actual attitudes toward social practices might be unreasonably influenced by the benefit they receive from the practice. Hence, a lack of regret among some of those identified as having been wronged by a moral principle does not seem to be grounds for disqualifying the principle that holds them to have been wronged.

Because the antipoverty principle does not require comparisons to other possible states of affairs in which identical persons would have existed, it is not undermined by the no-worse-off argument. Moreover, it is not obviously irrational for persons to judge that the institutional order to which they owe their existence is unjust. And even if we suppose that they cannot wish not to have existed, this does not provide a reason to dismiss the antipoverty principle. Finally, it does not seem to be the case that the absence of regret on the part of those who have allegedly been wronged renders the alleged wrongdoing not wrong. Therefore, the non-identity problem does not pose great problems for the antipoverty principle.

Climate Change and the Human Rights of Future Persons: Assessing Four Philosophical Challenges

In this appendix I examine a model argument, invoking human rights in defense of the duties to future persons to mitigate – and provide adaptation to – climate change. I look at four challenges to the human rights argument. The first three have been pressed in the literature on conceptual grounds. The fourth I formulate on normative grounds. I present what I think are satisfactory responses to the three conceptual challenges. The fourth challenge shows, I think, the limits of the human rights approach for guiding policy. In defending that view I assume that there are important moral duties to mitigate climate change and to provide adaptation assistance, and I take human rights seriously. The issue is whether they alone can guide us in making the kinds of trade-offs that climate change policy seems to require. The antipoverty principle can, I think, do that job.

THE HUMAN RIGHTS ARGUMENT

In the absence of additional mitigation, climate change is projected to have profound, often devastating, effects on hundreds of millions of people by the end of this century. Much of this has been discussed in the chapters of this book, but there are some reminders relevant to this appendix. First, human health is expected to suffer significantly. According to the Intergovernmental Panel on Climate Change (IPCC), "The health status of millions of people is projected to be affected through, for example, increases in malnutrition; increased deaths, diseases and injury due to extreme weather events; increased burden of diarrheal disease; increased frequency of cardio-respiratory diseases . . . and the altered spatial distribution of some infectious diseases."[1] Second, for hundreds of millions of people, access to water and food will become more difficult.

[1] IPCC, 2007 Synthesis, p. 48.

By 2020, anywhere from 75 to 250 million Africans are expected to suffer increased water stress, and yields on rain-fed farms may be reduced by up to 50 percent.[2] According to a United Nations Human Development Programme review of climate change projections, "Overall, climate change will lower the incomes and reduce the opportunities of vulnerable populations. By 2080, the number of people at risk of hunger could reach 600 million – twice the number of people living in poverty in sub-Saharan Africa today."[3]

Third, millions of people are already at risk of extreme weather and flooding. Currently around 344 million people are exposed to tropical cyclones, 521 million to floods, 130 million to droughts, and 2.3 million to landslides.[4] Climate change is expected to increase these numbers very significantly and the poor are especially vulnerable. "People living in the Ganges Delta and lower Manhattan share flood risks associated with rising seas levels. They do not share the same vulnerabilities. The reason: The Ganges Delta is marked by high levels of poverty and low levels of infrastructural protection."[5] About 10 percent of the world's population lives at an elevation of 10 m or less above sea level.[6] Hundreds of millions of people, then, are at risk of inundation from floods, storms, and rising sea levels. But the poor living in the mega deltas of North Africa and Asia are particularly vulnerable. The devastation caused by drought and flooding could result in long-term setbacks to human development in poor societies.[7]

In light of the calamities forecasted to be caused by climate change, the protections offered by international human rights documents seem important. For example, Article 25, paragraph 1 of the Universal Declaration of Human Rights states that "[e]veryone has the right to a standard of living adequate for the health and well-being of himself and of his family, including food, clothing, housing and medical care and necessary social services, and the right to security in the event of unemployment, sickness, disability, widowhood, old age or other lack of livelihood in circumstances beyond his control."[8] Article 11 of International Covenant on Economic, Social and Cultural Rights holds that

---

[2] *Ibid.*, p. 50.

[3] United Nations Human Development Programme, *Human Development Report 2007–2008*, p. 90.

[4] *Ibid.*, p. 98.

[5] *Ibid.*, p. 78.

[6] Gordon McGranahan, Deborah Balk, and Bridget Anderson, "The Rising Tide: Assessing the Risks of Climate Change and Human Settlements in Low Elevation Coastal Zone," *Environment and Urbanization* 19 (2007): 17–37.

[7] UNHDP, *Human Development Report 2007–2008*, pp. 88–89.

[8] Universal Declaration of Human Rights, Art. 25, para. 1. http://www.un.org/en/documents/udhr/index.shtml (accessed December 26, 2012).

"[t]he States Parties to the present Covenant recognize the right of everyone to an adequate standard of living for himself and his family, including adequate food, clothing and housing, and to the continuous improvement of living conditions."[9]

The human rights approach has been taken up by several philosophers. For example, Simon Caney argues that a central and fundamental wrong of climate change is that it will cause significant human rights violations.[10] Caney makes this argument in relation to three key rights: (1) The right not to be arbitrarily deprived of one's life; (2) the right not to have others cause serious threats to one's health; and (3) the right not to have others deprive one of the means of subsistence. With respect to the first right, climate change-caused death by starvation, dehydration, tropical diseases, and flooding appear to be instances of the arbitrary deprivation of one's life. Insofar as those who die are neither found to be guilty in a criminal process nor are they combatants in war, their deaths are arbitrary. Moreover, insofar as the deaths are effects of greenhouse gas emissions, they are plausibly deprivations and not merely the outcome of a failure to rescue those who are in need. With respect to the second and third rights, the reasoning is similar: hunger and dehydration (as well as disease) pose serious threats to human health and are usually instances of a lack of means of subsistence. Insofar as they are caused by greenhouse gas emissions, they satisfy the second right's requirement of being human caused and the third's of being deprivations.

It is important that these three rights are not controversial. They are both less demanding and less controversial than the rights enumerated in the paragraphs of the human rights documents cited earlier. Hence, Caney argues that these three claims are especially good candidates for rights claims that can guide climate change policy. The model argument goes like this:

1. The human rights of future peoples should not be violated, instances of which rights include: (1) being arbitrarily deprived of one's life, (2) suffering serious health threats, and (3) being deprived of the means of subsistence.

---

9 International Covenant on Economic, Social and Cultural Rights, Article 11, para. 1. http://www2.ohchr.org/english/law/cescr.htm (accessed December 26, 2012).
10 Simon Caney, "Climate Change, Human Rights, and Moral Thresholds," in Stephen Humphreys, ed., *Human Rights and Climate Change* (Cambridge: Cambridge University Press, 2010), pp. 69–90. David Miller endorses the importance of human rights to moral assessment of climate change in David Miller, "Global Justice and Climate Change: How Should Responsibilities Be Distributed?" The Tanner Lectures on Human Values, Tsinghua University, Beijing, March 24–25, 2008. http://www.tannerlectures.utah.edu/lectures/documents/Miller_08.pdf (accessed December 26, 2012).

2. Unless the present generation pursues an appropriate policy-combination of reductions of present and future greenhouse gas emissions and adaptation to anthropogenic climate change, historical and present greenhouse gases will bring about instances of deprivation of the three sorts listed earlier.
3. Therefore, unless the present generation pursues an appropriate policy-combination of reductions of present and future greenhouse gas emissions and adaptation to anthropogenic climate change, its policies will violate duties it has to future people because of their human rights.

For the sake of convenience, I refer to this as *the human rights argument for mitigation and adaptation* or often simply as *the human rights argument*. By doing so, of course, I do not mean to assume that there is only one human rights argument; it does seem, however, to be an especially perspicuous model.

THE HUMAN RIGHTS ARGUMENT AS INTERGENERATIONAL JUSTICE

The first premise invokes the human rights of future persons. In so doing, it situates the argument in discussions about intergenerational justice. The moral duty to pursue policies of mitigation and adaptation strategies in the conclusion is then a duty of intergenerational justice. In light of this, it might be objected that the argument ignores present suffering caused by anthropogenic climate change. The global mean temperature has already increased by approximately 0.74°C.[11] Surely, this increase is already destabilizing weather patterns, resulting in heat waves, droughts, and more frequent storms. The best explanation for the warming is anthropogenic greenhouse gas emissions. Therefore, there is a human stamp on the suffering that is currently occurring around the world owing to heat waves, droughts, and floods.

The response to this objection, I think, should be to grant that the human rights argument, as it is stated here, does not offer the reasons to eradicate current suffering caused by climate change. This is not to assert that there are no such reasons. Outcome responsibility is crucial to Caney's account (see premise two), and the greenhouse gas emissions policies of the present generation is not outcome responsible (or not very responsible) for the suffering of the present generation. Stocks of CO_2 in the atmosphere are the accumulation of well more than 100 years of emissions. Although there has been a rapid increase in emissions in the last twenty years or so, most people now living are

---

[11] IPCC, 2007 Synthesis, p. 30.

younger than thirty years old and therefore not plausibly outcome responsible for the bulk of the emissions causing current misery.

Nothing prevents augmenting the human rights argument with other considerations of global justice or humanity to provide moral reasons to eradicate current suffering. It might be best in any case to discuss the reasons to eradicate such suffering within the framework of global justice or humanitarian duties rather than climate change–related justice because the claim that the cause of a particular instance of suffering is caused by climate change is usually based on weak scientific ground. When it comes to the climatic effects of global warming, the best that the present state of climate science can do is to argue in terms of general trends and frequencies. Heat waves, droughts, floods, and tropical storms occurred before the increase in mean global temperature. They are also the result of non-anthropogenic factors. Given the current state of the science, one cannot state with a high degree of confidence that any particular weather event or cycle is the product of merely anthropogenic factors, or even what proportion of the factors are anthropogenic. So, it is best to press the case for eradicating present suffering on grounds other than those related to demands of justice on our generation to mitigate and adapt to climate change.

THE FIRST CHALLENGE (CONCEPTUAL): THE NONEXISTENCE OF FUTURE PERSONS

Arguments like the human rights argument for mitigation and adaptation have been challenged by Wilfred Beckerman and Joanna Pasnek on grounds that people who do not exist can have no rights and therefore we can have no rights-based duties toward them.[12] Their argument could be pressed by means of showing an implausible implication of claiming that nonexistent future people currently have rights. The following is an example of such an argument.

Population Control

If future people currently have any human rights at all, they currently have the right to life. But if they currently have the right to life, then no population control policy put in place now could be justified because it would prevent

---

[12] See Wilfred Beckerman and Joanna Pasek, *Justice Posterity, and the Environment* (Oxford: Oxford University Press, 2001), pp. 15–23.

future people from coming into existence. Given that at least some population policies must be justified, future people do not currently have the right to life, or any other human rights.

One might want to respond to Population Control by criticizing its soundness; in particular, some might maintain that it is possible for people to have human rights, but not the right to life. Doing so would allow one to make the human rights argument on grounds (2) and (3) – the right not to have others cause serious threats to one's health and the right not to have others deprive one of the means of subsistence – but not ground (1) – the right not to be deprived of one' life. However, this seems like an unpromising response given that it is hard to see why if one can arbitrarily deprive another of his life without violating his rights, anything else that one did to him would be a violation of his rights.

An alternative response to Population Control would be to deny that preventing someone from existing is a violation of the person's right to life. This would avoid the problems of the first response by preserving the idea that future people currently have the full package of human rights, including the right to life. However, it would deny that the latter entails that one may not prevent another from existing. The idea would be that the right to life proscribes only arbitrary killing, not the prevention of coming into existence. This would be to allow criticizing energy policies on the grounds that they result in arbitrary deprivations of life, but not criticizing population control policies on the same grounds.

The idea behind this second response would be that the point of the right to life is to proscribe certain actions that move people from existing to not existing (in other words, killing), but not to proscribe policies that prevent people from moving from not existing to existing. If that is the case, however, the right to life would not be parallel to the other two categories of human rights mentioned by Caney. The rights not to be deprived of health and the means of subsistence both seem to prevent certain actions that move people away from health and being in possession of the means of subsistence and to proscribe policies that prevent people from moving from poor health to health and from lacking the means of subsistence to having them. Generally, human rights proscribe both preventing people from the enjoyment of the right and dispossessing people of their enjoyment of it. Perhaps the right to life is the exception, but it is not clear why that would be the case.

The best response to Population Control is, I think, simply to accept it. Future people do not currently have any human rights. When they come to exist they will, but it is important to appreciate that this latter claim is sufficient to place current people under duties to future people. We can be under a duty not to act in a way that will bring about a condition in which people will have rights that we either cannot or do not intend to fulfill, as revealed by thinking about why we should not make promises we do not intend to keep.

Lying Promise

Generally, the promisee has a defeasible moral right to expect the promisor to perform as promised. But the would-be promisee has no right concerning the same kind of action by the would-be promisor in the absence of the promise. It does not follow, however, that the would-be promisor has no promise-related duties to the would-be promisee. If he makes a promise, he will have the duties of a promisor. But even in the absence of such duties, he has a duty not to create a right held by the would-be promisee that he (the would-be promisor) does not intend to, or cannot, keep. The rights that the promisee will have once a promise is made (but does not have as merely a would-be promisee) constrain the would-be promisor.

One should not make promises one does not intend to keep because once the promise is made, the person to whom the promise is made has a moral right to have the promise fulfilled. The duty not to make a promise one does not intend to keep is not derived from the existing right that the promisee has that the content of the promise be fulfilled. Rather, it is based on the right the promisee would have if one promised. The possible future right is the source of a present duty.[13]

The human rights argument for mitigation and adaptation can be seen as arguing analogously for the duties that members of present generations have in light of the human rights of future people. This entails that the conclusion concerning what is wrong with policies that fail to mitigate or to provide for adaptation must be carefully stated. The promisor does not break a promise to a promisee by making a promise he cannot or intends not to keep. The wrong, rather, is in making the promise, which has not yet been broken. Similarly, the wrong of policies that fail to mitigate or to provide for adaptation is not that they violate the current rights of future people; they have no such rights.

---

[13] For a different argument with approximately the same conclusion, see Derek Bell, "Does Anthropogenic Climate Change Violate Human Rights?" *Critical Review of International Social and Political Philosophy* 14 (2011): 104–110. Simon Caney also criticizes Beckerman and Pasek in his "Human Rights, Responsibilities, and Climate Change," in Charles R. Betiz and Robert E. Goodin, eds., *Global Basic Rights* (Oxford: Oxford University Press, 2009), pp. 234–236.

The wrong is that such policies make it impossible to honor the rights that persons in the future will possess. This explains why the conclusion of the human rights argument does not claim that failure to mitigate or to provide for adaptation just is a human rights violation.

THE SECOND CHALLENGE (CONCEPTUAL): WAIVER AND THE CONTINGENT EXISTENCE OF FUTURE PERSONS

Two arguments against invoking the rights of future people presented by Derek Parfit in his discussion of the non-identity problem pertain to the contingency (rather than simply the temporal location) of future persons.[14] Neither argument denies that future people will have rights. But they both deny that it is coherent to invoke such rights to account for the duties of people in the present. One argument employs the idea of a waiver.

Waiver

A rights-holder has it in her power to waive her rights. Her rights are claims on others that they not act in a way that would violate a right of hers. If she waives a right, then an action by another that in all other respects is of the kind that would ordinarily be a rights-violation is not a violation. For example, if someone taken into custody by the police were to waive her right to a trial before being found guilty and sentenced, then the authorities would be released of their duties to try her before pronouncing her guilty and sentencing her.

The relevance of Waiver to climate change derives from the contingency of the existence of future people. A comprehensive energy policy has profound and widespread effects on how, how often, and how far people travel, on how they sustain a livelihood, on where they live, and as a result of all of these factors, also on whom they meet in the course of their lives. People will meet who would not have met, given a different energy policy. Insofar as a person's identity is contingent on whom his parents are and when he is conceived, over time, different people will come into existence given different comprehensive energy policies. Suppose, owing to a failure to mitigate climate change sufficiently, people come into existence under conditions that, say, threaten their enjoyment of their rights not to be deprived of their health and means of subsistence. Upon reflecting on the contingency of their existence – in particular, that they would not have existed under a comprehensive energy

---

[14] Derek Parfit, *Reasons and Persons* (Oxford: Oxford University Press, 1987), chp. 16.

policy that adequately mitigated climate change – if they deem their life worth living, they might waive their rights not to be deprived of their health or means of subsistence.[15] In that case, the failure to mitigate did not produce a human rights violation.

There is a disanalogy between Waiver and the climate change case, however. In Waiver, the release occurs before what would otherwise be a rights violation. In the climate change case, the release is retrospective. Hence, in the example imagined in Waiver, the guilty verdict and sentencing proceeds confidently on the basis that there is no rights violation. In the case of failing to mitigate climate change, policy makers have assurance of the permissibility of their action only to the extent that they have assurance that future people will waive their rights. This raises the question of whether an actual waiver or a hypothetical waiver is required. If an actual waiver is required, there can be no assurance that the failure to mitigate will not constitute a rights violation because we cannot know whether people will waive their rights. In order to undermine the human rights argument for mitigation and adaptation, it must be the case that the waiver necessarily would be forthcoming in a hypothetical choice situation, regardless of whether it is actually offered. It must be the case either that fully rational people reflecting on their interests would waive their rights not to be deprived of health and the means of subsistence or that people reflecting in a way that models what is reasonable to demand of others would waive these rights.

An argument that a fully rational person would waive his rights in such circumstances would go as follows:

Rational Waiver

Not to waive my rights would be to will that the past would have been different than it is. As a part of practical reason, rational willing extends like a ray from the present into the future. Although we might seek to understand the past, we cannot will it to have been otherwise. Hence, I must waive my rights.

One problem with Rational Waiver is that seeking enforcement of rights is often less about seeking to change the past than about seeking remedies in the present. In an important response to Parfit, James Woodward offers the example of a person denied boarding on a flight on racist grounds, who, even though his life is spared because the plane crashes and kills everyone onboard, pursues a suit against the airline for discrimination.[16] Pressing the suit should

---

[15] Parfit, *Reasons and Persons*, p. 364.
[16] James Woodward, "The Non-Identity Problem," *Ethics* 96 (1986): 810–811.

not be understood as irrationally trying to change the past, but seeking remedy for a wrong done. The defender of Rational Waiver might reply that it makes little sense for a person to press for a remedy from others who are long dead. But this may or may not be the case depending on what has become of the wealth of the dead people. Moreover, there is something else that the living might rationally want – namely, public acknowledgment that they have been wronged. They cannot get such acknowledgment and waive their rights. So, it is by no means obvious that fully rational people would necessarily waive their rights after reflecting on the fact that they would not exist but for the policies that caused their rights to be violated.

Let's consider the possibility, then, that it would be unreasonable for someone not to waive his rights.

Reasonable Waiver

> To fail to waive my rights in the circumstances is to demand more than I am entitled to from previous generations. It is to demand both my existence and the fulfillment of my rights. At most I am entitled to one or the other, but not both, of these.

A sufficient response to Reasonable Waiver is to deny that failing to waive one's rights entails demanding both one's existence and one's rights. Failing to waive one's rights might be based not on the conjunction, but on a conditional claim: If I am to be given an existence, it ought to be under conditions in which my rights are not deprived. The energy policy of the previous generation failed to satisfy that condition and it is therefore wrong. This is consistent with the argument of the first challenge, which took the wrong of a failure to mitigate to be that it brought into existence people whose rights could not be respected.

In sum, neither Rational Waiver nor Reasonable Waiver is compelling. Hence, there would appear to be no incoherence – on grounds that rights might be waived – to the conclusion of the human rights argument.

THE THIRD CHALLENGE (CONCEPTUAL): IMPOSSIBILITY AND THE CONTINGENT EXISTENCE OF FUTURE PERSONS

A second argument that Parfit makes against invoking the rights of persons whose existence is in part contingent on the action or policy that putatively violates the rights is based on the apparent impossibility of respecting the rights in question.[17]

---

[17] Parfit, *Reasons and Persons*, p. 365.

Impossibility

Because the existence of future people is contingent on the pursuit of a policy that fails to mitigate climate change, it is impossible to respect their rights not to be deprived of health and the means of subsistence. Since "ought" implies "can," by contraposition, there is no duty to respect these rights.

In response, if future people whose lives would be contingent on our failure to mitigate currently had the right to life, which required bringing it about that they exist, then it would be impossible to respect that right and their rights not to be deprived of health and the means of subsistence. But then we would be in a colossally tragic situation in which it would be impossible to respect the rights claims of all possible future people given that different people will be brought into existence by different policies and we obviously cannot pursue all possible policies. But as I argued in response to the first challenge, taking ourselves to be under duties because of the rights of future persons (contingent or otherwise) does not require assuming that they presently have those rights.

An adequate response to Impossibility is simply to deny the premise that it is impossible to respect the rights of future persons. They do not have those rights now, and therefore they have no right to be brought into existence. Instead, we respect them by not bringing them into existence in conditions in which it would be impossible for some of their rights to be respected.

The past three sections have demonstrated, I think, that there is nothing incoherent about claiming that the rights of future people can be the source of duties to present people. If so, then the human rights argument for mitigation and adaptation is not (at least for the reasons invoked by these arguments) fundamentally incoherent.

THE FOURTH (NORMATIVE) CHALLENGE: TRADE-OFFS

The threat that climate change will produce massive human suffering and negatively affect vital human interests is real. This is what motivates moral concern about the condition of future people. But climate change mitigation policies also pose threats. Indeed it is not possible to understand the current impasse in international negotiations without appreciating that mitigation also poses a threat. Suppose that an adequate mitigation policy would result in a 2 percent reduction in global gross domestic (GDP) product by 2030.[18] Developing and underdeveloped countries worry that the loss in GDP will be assigned

---

[18] The estimate comes from Nicholas Stern. See Nicholas Stern, *The Global Deal: Climate Change and the Creation of a New Era of Progress and Prosperity* (New York: Public Affairs,

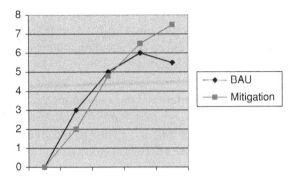

FIGURE B.1. Two possible intergenerational growth curves.

in a way that makes it nearly impossible for them to make any progress in human development. Human development is tremendously energy intensive. If an increase in the price of carbon is uniformly applied and enforced globally, or if emissions reductions are uniformly applied and enforced without wealth or technology transfers, it will be nearly impossible for many developing and underdeveloped countries to make substantial progress in human development.

The argument often given by economists to justify accepting mitigation costs of approximately 2 percent of global GDP is that there will be even greater GDP reductions in the future unless the mitigation costs are assumed. Figure B.1 is meant to represent these cost forecasts in a general way, without an attempt to provide accurate numbers. The vertical axis represents global GDP; the horizontal axis represents time. At some point in the future, policies that have mitigated climate change earlier produce a higher GDP than policies that did not pursue mitigation. This higher GDP continues indefinitely, thereby providing the economic rationale for accepting mitigation.

According to the economic argument, the moral concern about the effects of climate change is a concern, in part, about what happens to the right of the intersection of the two curves representing economic growth on a version of business-as-usual – that is, without mitigation – and economic growth with mitigation. The costs of climate change are represented by the area between the higher Mitigation line and the lower BAU line. For ease of discussion, let's refer to this as *the right area*. Figure B.1 also makes lucid that achieving adequate mitigation will incur costs. These are represented by the area to the

2009), p. 51. For reasons discussed in Chapter 4, we cannot simply take Stern's account of the costs at face value – be we needn't worry about that here.

left of the intersection, where the BAU curve is higher and the Mitigation curve is lower. Let's call this *the left area*.

The human rights argument directs our attention to the human rights violations that are associated with the costs of the right area. The idea is that costs will be assigned in such a way that people will die, become ill, and/or lose their means of subsistence. Given current trends in global inequality and poverty, that seems reasonable. The concern here is not about the economic costs of climate change. If we could proceed on the BAU path with respect to global warming and its attendant climate effects and redistribute global resources in such a way that the poor and the especially vulnerable (typically, classes of people that overlap considerably) were sufficiently insured and provided with means of adaptation, then the costs represented by the right area would not provide reasons for moral condemnation.

Similar considerations apply to the left area. Given current trends in global inequality and poverty, the costs of mitigating climate change might seem most likely to fall on the poor with the predictable effect that their rights – not to be arbitrarily deprived of life, health, and the means of subsistence – will be violated. Suppose that that pessimistic prediction about climate change mitigation were correct; in that case, it would be unclear how the human rights approach gives a reason to prefer Mitigation over BAU given that massive human rights violations would occur either way, along BAU to the right of the intersection point and along Mitigation to the left.

Providing guidance for policy in the case of competing human rights claims could perhaps be accomplished if we are to pursue the course of action that is likely to lead to maximal satisfaction of human rights. We notice that the right area is much larger than the left area and we suppose that there will be many more human rights violations there too (on the assumption that the distribution of the economic costs produces human rights violations). Therefore, a consequentialism of rights recommends Mitigation.

There are respectable philosophical accounts of rights that take them as the objects of maximizing strategies.[19] However, an important part of the moral importance of human rights is to offer protections to minorities against the demands of majorities, and it is at least questionable that a consequentialism of rights can adequately accomplish that. If our account of human rights allows

---

[19] For a maximizing approach to rights, see Amartya Sen, "Rights and Agency," *Philosophy and Public Affairs* 11 (1982): 3–39. For deontological approaches including rights as trumps, see Ronald Dworkin, *Taking Rights Seriously* (Cambridge, MA: Harvard University Press, 1978); rights as side constraints, see Robert Nozick, *Anarchy, State and Utopia* (New York: Basic Books, 1977); and rights as protectors of important interests, see Joseph Raz, *The Morality of Freedom* (Oxford: Oxford University Press, 1986).

us to favor the right area over the left, then why not prefer the democratic rights of the majority over the rights of conscience of a religious minority, which the majority seeks to suppress? My purpose here is not to resolve the debate within human rights theory about whether a consequentialism of rights is the best approach, but to note a problem for it and to suggest an alternative if the problem runs as deep as it seems.

Another response to the problem of competing human rights claims is to deny the pessimistic prediction about climate change mitigation and therefore the need to maximize the fulfilment of rights. Given the veto power of the least-developed and developing countries in international climate change negotiations, it seems doubtful that a mitigation treaty will make human development so costly as to be out of reach for many countries.[20] It is more likely that there will be no serious mitigation treaty. Although this response offers reasons to believe that the costs of a mitigation treaty will not make worse the condition of vulnerable people living in the developing and underdeveloped world, it does not ensure that the costs of mitigation will not fall on the already vulnerable, wherever they might live. If Mitigation raises the price of energy, many poor and relatively immobile people living in developed countries may be at greater risk of death owing to a lack of heating and air conditioning. The problem of competing rights claims seems, then, to persist even if the political reality is that negotiators from the developing world will not let their countries be bullied into a treaty that prevents human development.

One final response seems available to the defender of the human rights argument: Policy properly guided by the human rights argument would not be responsible for any human rights violations either to the left or to the right of the intersection point. This claim seems to have a heroic picture of mitigation policy in mind, not unlike the following one.

Policy Control Tower

Somewhere there is a policy control tower, large enough to contain all possible policy levers, and high enough to foresee all possible consequences. It is left to us only to take our seats in the tower and to pull all necessary levers, each in the right sequential order, so that the result is that the costs of climate change mitigation are distributed in such a way that no one's human rights are violated. The result is a distribution of costs along the Mitigation curve without any infringement of human rights.

---

[20] I also defend this view in Darrel Moellendorf, "Treaty Norms and Climate Change Mitigation," *Ethics and International Affairs* 23 (2009): 262.

Anyone defending using human rights to guide climate change policy needs a plausible story for why the person in the tower could not also pull levers resulting in no human rights violations as the world economy is steered along the BAU curve. Perhaps this cannot be done because at some point the curve will drop off so sharply to the right that no amount of lever pulling will prevent human rights violations. But the plausibility (such as it is) of the policy control tower example rests on the assumption that the curves in the graph necessarily do not require human rights violations; any such violations are merely contingent and can be avoided by intelligent lever-pulling strategies. If that assumption is true, then all the right levers, in the right order, could in principle be pulled for BAU as well as Mitigation. Now, if that is implausible for BAU, then it is not clear why it can be assumed for Mitigation.

The relationship between both curves and human rights violations seems to be an empirically contingent matter. And, it is doubtful that we know enough to suppose that all the right policy levers, in the right order, could even in principle be pulled along the Mitigation curve. We have a lot to learn about climate change mitigation policy before we will get it right. We are likely to make many mistakes along the way. The difficulties that the European Union's Emission Trading Scheme has had in putting the right price on emissions permits are a reminder of this. We do not have the user's guide to the policy control tower, even if it contains all the right levers. More importantly, it is implausible to claim that all the right levers are available. International climate change policy will come to pass as a result of international negotiations. In international negotiating forums, the current strength of the norm of state sovereignty prevents ensuring that states not assign the costs of their mitigation obligations in ways that produce human rights violations.

Insofar as the policy control tower example assumes that *in principle* all the right policy levers, in the right order, could be pulled to avoid human rights violations, it is at best a piece of ideal theory. But to the extent that it is ideal theory, it is not clear what sort of guidance it offers in approximating the ideal. Climate change policy will inevitably involve considerations of trade-offs. There seems no way to avoid comparing harms. Insofar as that is the case, morally sensitive policy needs an account of how to make trade-offs. Appealing to human rights provides insufficient guidance with respect to that.

CONCLUSION

I have surveyed three arguments in the literature that assert that there is a conceptual confusion involved in using human rights to talk about the moral claims of future people. I don't think that these arguments are convincing and

I have tried to say why. I also sketched a normative argument to the conclusion that the model human rights argument for mitigation and adaptation fails to provide sufficient guidance in making the trade-offs that seems a part of climate change policy. The limitations of the human rights argument suggested by this criticism point to the need for a moral principle that allows making judgments that involve comparing policies that contain bad outcomes and avoiding the ones, which under the guidance of a compelling moral principle, we have the best reasons to avoid. I argued in Chapter 1 that the antipoverty principle can do that.

The Right to Sustainable Development versus International Paretianism

Eric A. Posner and David Weisbach's central argument against the right to sustainable development is perhaps charitably made fully explicit as follows: (1) States act (almost) only on their perceived interests. Therefore, (2) a policy is feasible only if it is in the perceived interests of all of the states making the policy. (3) Policy pursuing the right to sustainable development necessarily involves international redistribution. (4) International redistribution is not in the perceived interests of states out of which resources would flow. Therefore, (5) a policy pursuing the right to sustainable development is infeasible. However, because this does not establish that international Paretianism is feasible, more must be said to avoid the tragic situation in which there are no feasible policies. Thus, the argument continues. (6) International Paretianism is in the perceived interests of all states making policy. Therefore, (7) international Paretianism is feasible. In this appendix I shall discuss three problems that render this argument unsound.

The most significant problem with the argument is logical. It does not rule out the possibility of the tragic situation; this is because the inference to (7) is invalid. To see this, focus your attention on premises (2) and (6). Premise (2) states what must be the case if a policy is to be feasible. In other words, it states a *necessary condition* of feasibility. Premise (6) states that international Paretianism meets the necessary condition. It does not follow that international Paretianism is feasible.

This is like concluding that John understands a book from the premises that John only understands books written in English and the book is written in English. Books might need to have certain other features as well, such as clarity of style and argument, to be understood by John. To get to the conclusion, we need a premise that John understands all books in English. For if there are no books in English that John does not understand (not even this one!), and the book is in English, then one can conclude that John understands the book.

Back to Posner and Weisbach's argument. To make the argument to (7) valid, premise (2) would need to be replaced with one stating a *sufficient condition*. The argument requires the following premise: (2*) A policy is feasible if it is in the perceived interests of all of the states making the policy. But the logical problem with substituting (2*) for (2) is that (2*), (3), and (4) do not establish the sub-conclusion (5), which rules out policies that pursue the right to sustainable development on feasibility grounds. In order to rule out the right to sustainable development, one has to state a necessary condition and then claim that the condition is not satisfied. And (2*) does not state a necessary condition.

The logic of the argument can be salvaged if (2) is replaced with a premise that states both a necessary and a sufficient condition, such as (2**) A policy is feasible if and only if it is in the perceived interests of all of the states making the policy.

Is premise (2**) plausible? We must ask what is meant by the claim that a policy is in the perceived interest of a state. For Posner and Weisbach, the baseline is a business-as-usual scenario. For simplicity's sake, suppose that there are two states (*Rich* and *Poor*). The business-as-usual is option (i). The only choice is whether to depart from business-as-usual to either one of the following two options, both of which achieve sufficient mitigation: (ii) a perceived improvement for both *Rich* and *Poor* in comparison to business-as-usual; or (iii) a perceived improvement for both in comparison to business-as-usual, but less for *Rich* and more for *Poor* than in (ii). From among the three, (ii) is in the perceived best interests of *Rich*. Meanwhile (iii) is in the perceived best interest of *Poor*. However, according to Posner and Weisbach's account of efficiency, both satisfy international Paretianism. So, (2**) would rule out (i), but would be neutral between (ii) and (iii).

But suppose that *Rich* is a highly developed state and *Poor* is a developing state containing extensive poverty. And suppose further that under (ii), *Poor* expects to remain underdeveloped for the foreseeable future, but under (iii), it expects advancement. In other words, in (iii) *Poor* expects to enjoy the right to sustainable development, but not in (ii). Then I submit that premise (2**) would be implausible given the following reason of realism: A state will not agree to a policy that it expects to forestall its economic development if there is an alternative that does not.[1] The inclusion of the sufficient condition in (2**) renders it false. Premise (2) would not be false given the

---

[1] Henry Shue made this point some time ago in "The Unavoidability of Justice," *The International Politics of the Environment: Actors, Interests, and Institutions*, Andrew Hurrel and Benedict Kingsbury, eds. (Oxford: Oxford University Press, 1992), pp. 377–378.

ground of realism, but this simply returns us to the logical problems of the argument. In short, the argument is either invalid or contains an implausible premise.

The second problem with the argument on any of these three formulations concerns the plausibility of premise (3) – namely, that policy pursuing the right to sustainable development involves redistribution. Redistribution involves causing a transfer from one party to another. Suppose the same scenario as in the previous two paragraphs, but where the choice is between different actual outcomes and not simply between what states believe about the world. When comparing (ii) to (iii), it might seem like (iii) is redistributive since *Rich* would do better in (ii) and *Poor* better in (iii). But the appearance is misleading because there is no way to cause a transfer from *Rich* to *Poor* that would produce a change from (ii) to (iii). Rather, the choice is between (ii) and (iii) as departures from (i), business-as-usual. Because everyone is better off, the improved circumstances of *Rich* and *Poor* in (ii) and (iii) are the results of growth, not of redistribution. Choice (iii) secures the right to sustainable development, then, without redistribution. Securing the right to sustainable development is not necessarily redistributive. And it certainly isn't if the circumstance in which the right is secured is Pareto Superior (not merely believed to be so) to business-as-usual.

It should be stressed that in fact it is quite plausible that securing the right to sustainable development will also be internationally Pareto Superior to business-as-usual. Recall that business-as-usual is likely to bring about warming well in excess of 4°C. This is likely to produce significant hardships in the form of flooding, droughts, heat waves, and tropical storms even in the highly developed world. It seems very plausible that in comparison to business-as-usual, highly developed states will be better off under an international mitigation treaty that recognizes the right to sustainable development.

Premise (3), however, might envision a choice that does not correspond to the plausible picture just drawn. This is the kind of situation discussed in Chapter 5 in which from (i) the choice is between (ii) and (iv), where (iv) also sufficiently mitigates climate change, but in (iv) *Rich* is mildly worse off, but still highly developed, and *Poor* is no longer underdeveloped. In this case, the choice is between securing international Paretianism and the right to sustainable development. A move from (i) to (iv) might be the effect of a transfer of resources from *Rich* to *Poor*. Aid in the form of money might be given to *Poor* by *Rich*; technology might be provided; and intellectual property might be shared. But the move from (i) to (iv) also might be owing to different stricter regulations regarding carbon emission in *Rich* than in *Poor*. These different regulations affect the growth trajectories of the two states, but they do not cause a transfer of resources. Once again, securing the right to sustainable development is not necessarily redistributive.

If (3) is false, then even if the argument from (2), (3), and (4) to sub-conclusion (5) is valid, it is nonetheless unsound.

The third problem with the argument on any of the formulations I have discussed concerns the truth and importance of (1). The term "almost" is the problem. It is there because this seems to be what Posner and Weisbach believe. For they claim that "history provides very few cases where states act against their own perceived interests in order to satisfy the moral claims of other states."[2] But the presence of "almost" destroys the inference from (1) to (2). It therefore also calls into question the soundness of the argument from (2), (3), and (4) to sub-conclusion (5) given that it is possible that recognizing the right to sustainable development would be another one of those "very few cases." The alternative is to excise "almost" and to construe the argument in a way that seems unintended and is, perhaps, factually inaccurate. If the speculations entertained in discussing the second problem are plausible – that is, if a mitigation regime that recognizes the right to sustainable development would be internationally Pareto Superior to business-as-usual – then there is not much need to worry about whether states ever act contrary to their interests. We are, in that case, discussing about an utterly trivial matter.

But at the risk of spending a little time on a trivial matter, let's suppose that a mitigation regime that recognized the right to sustainable development were not in the perceived interests of the highly developed states. Might climate change be the kind of problem in which the "almost" would peek out of the parentheses and do the work of undermining the inference to the conclusion that the right to sustainable development is infeasible? If the aforementioned reason of realism is compelling – namely, a state will not agree to a policy that it expects to forestall its economic development if there is an alternative that does not – then there might be significant pressure from within the populations of highly developed states to take the moral claims of least-developed and developing states seriously and not force them into underdevelopment by means of either climate change or a climate change regime. The moral reasons that motivate such pressure might be augmented by a moral concern about future generations that is itself augmented by the prudential consideration that the calculation of the highly developed states could also be wrong. They might be making the world worse off for themselves as well. Such pressure would seem the best hope for progress.

---

[2] Eric A. Posner and David Weisbach, *Climate Change Justice* (Princeton, NJ: Princeton University Press, 2010), p. 6.

Declaration on Climate Justice

"All human beings are born free and equal in dignity and rights."
Article 1, The Universal Declaration of Human Rights

OUR VISION

As a diverse group of concerned world citizens and advocates, we stand in defense of a global climate system that is safe for all of humanity. We demand a world where our children and future generations are assured of fair and just opportunities for social stability, employment, a healthy planet and prosperity.

We are united in the need for an urgent response to the climate crisis – a response informed by the current impacts of climate change and the science that points to the possibility of a global temperature increase of 4°C by the end of this century. The economic and social costs of climate impacts on people, their rights, their homes, their food security and the ecosystems on which they depend cannot be ignored any longer. Nor can we overlook the injustice faced by the poorest and most vulnerable who bear a disproportionate burden from the impacts of climate change.

This reality drives our vision of climate justice. It puts people at the centre and delivers results for the climate, for human rights, and for development. Our vision acknowledges the injustices caused by climate change and the responsibility of those who have caused it. It requires us to build a common future based on justice for those who are most vulnerable to the impacts of climate change and a just transition to a safe and secure society and planet for everyone.

ACHIEVING CLIMATE JUSTICE

A greater imagination of the possible is vital to achieve a just and sustainable world. The priority pathways to achieve climate justice are:

Giving voice: The world cannot respond adequately to climate change unless people and communities are at the centre of decision-making at all levels – local, national and international. By sharing their knowledge, communities can take the lead in shaping effective solutions. We will only succeed if we give voice to those most affected, listen to their solutions, and empower them to act.

A new way to grow: There is a global limit to the carbon we can emit while maintaining a safe climate and it is essential that equitable ways to limit these emissions are achieved. Transforming our economic system to one based on low-carbon production and consumption can create inclusive sustainable development and reduce inequality. As a global community, we must innovate now to enable us to leave the majority of the remaining fossil fuel reserves in the ground – driving our transition to a climate resilient future.

To achieve a just transition, it is crucial that we invest in social protection, enhance worker's skills for redeployment in a low-carbon economy and promote access to sustainable development for all. Access to sustainable energy for the poorest is fundamental to making this transition fair and to achieving the right to development. Climate justice also means free worldwide access to breakthrough technologies for the transition to sustainability, for example, efficient organic solar panels and new chemical energy storage schemes.

Investing in the future: A new investment model is required to deal with the risks posed by climate change – now and in the future, so that intergenerational equity can be achieved. Policy certainty sends signals to invest in the right things. By avoiding investment in high-carbon assets that become obsolete, and prioritizing sustainable alternatives, we create a new investment model that builds capacity and resilience while lowering emissions.

Citizens are entitled to have a say in how their savings, such as pensions, are invested to achieve the climate future they want. It is critical that companies fulfil their social compact to invest in ways that benefit communities and the environment. Political leaders have to provide clear signals to business and investors that an equitable low-carbon economic future is the only sustainable option.

Commitment and accountability: Achieving climate justice requires that broader issues of inequality and weak governance are addressed both within countries and at a global level. Accountability is key. It is imperative that

Governments commit to bold action informed by science, and deliver on commitments made in the climate change regime to reduce emissions and provide climate finance, in particular for the most vulnerable countries.

All countries are part of the solution but developed countries must take the lead, followed by those less developed, but with the capacity to act. Climate justice increases the likelihood of strong commitments being made as all countries need to be treated fairly to play their part in a global deal. For many communities, including indigenous peoples around the world, adaptation to climate change is an urgent priority that has to be addressed much more assertively than before.

Rule of law: Climate change will exacerbate the vulnerability of urban and rural communities already suffering from unequal protection from the law. In the absence of adequate climate action there will be increased litigation by communities, companies and countries. International and national legal processes and systems will need to evolve and be used more imaginatively to ensure accountability and justice. Strong legal frameworks can provide certainty to ensure transparency, longevity, credibility and effective enforcement of climate and related policies.

TRANSFORMATIVE LEADERSHIP

World leaders have an opportunity and responsibility to demonstrate that they understand the urgency of the problem and the need to find equitable solutions now.

At the international level and through the United Nations, it is crucial that leaders focus attention on climate change as an issue of justice, global development and human security. By treating people and countries fairly, climate justice can help to deliver a strong, legally binding climate agreement in 2015. It is the responsibility of leaders to ensure that the post-2015 development agenda and the UNFCCC climate negotiations support each other to deliver a fair and ambitious global framework by the end of 2015. Local and national leaders will implement these policies on the ground, creating an understanding of the shared challenge amongst the citizens of the world and facilitating a transformation to a sustainable global society.

As part of global collective action, greater emphasis should be given to the role of diverse coalitions that are already emerging at the community, local, city, corporate and country levels and the vital role they play in mobilizing action.

These coalitions are already championing the solutions needed to solve the crisis and their effect can be maximized by supporting them to connect and scale up for greater impact.

Climate justice places people at its centre and focuses attention on rights, opportunities and fairness. For the sake of those affected by climate impacts now and in the future, we have no more time to waste. The 'fierce urgency of now' compels us to act.

This Declaration is supported by:

Mr Nnimmo Bassey, Coordinator, Oilwatch International

Ms Sharan Burrow, General Secretary, International Trade Union Confederation

Ms Luisa Dias Diogo, Former Prime Minister, Mozambique

Ms Patricia Espinosa-Cantellano, Ambassador of Mexico to Germany

Mr Bharrat Jagdeo, Roving Ambassador for the Three Basins Initiative, Former President of Guyana

Prof Pan Jiahua, Director General, Institute for Urban and Environmental Studies, Chinese Academy of Social Sciences

Prof Ravi Kanbur, TH Lee Professor of World Affairs, International Professor of Applied Economics and Management, Professor of Economics, Cornell University

Mr Caio Koch-Weser, Vice Chairman, Deutsche Bank Group, Chairman of the Board, European Climate Foundation

Mr Ricardo Lagos, President, Fundacion Democracia y Desarrollo, Former President of Chile

Mr Festus Mogae, Member, African Union High-level Panel for Egypt, Former President of Botswana

Mr Jay Naidoo, Chair of the Board of Directors, Chair of the Partnership Council, Global Alliance for Improved Nutrition

Mr Marvin Nala, Campaigner, Greenpeace East Asia

Prof Kirit Parikh, Chairman, Integrated Research and Action for Development, IRADe

Ms Sheela Patel, Founder-Director, Society for the Promotion of Area Resource Centers

Mrs Mary Robinson, President, Mary Robinson Foundation – Climate Justice, Former President of Ireland

Prof Hans Joachim Schellnhuber, Director, Potsdam Institute for Climate Impact Research

Prof Henry Shue, Senior Research Fellow at Merton College and Professor of Politics and International Relations, University of Oxford

Mr Tuiloma Neroni Slade, Secretary General, Pacific Islands Forum Secretariat

Dr Andrew Steer, President and CEO, World Resources Institute

Ms Victoria Tauli-Corpuz, Executive Director, Tebtebba Foundation

Ms Dessima Williams, Former Ambassador of Grenada to the United Nations

Those listed above are members of the High Level Advisory Committee (HLAC) to the Climate Justice Dialogue, an initiative of the Mary Robinson Foundation – Climate Justice and the World Resources Institute, that aims to mobilize political will and creative thinking to shape an ambitious and just international climate agreement in 2015.

Bibliography

Adorno, Theodor. *Aesthetic Theory*. Translated and edited by Robert Hullot-Kentor. London and New York: Continuum, 1997.

Allen, Myles R., David J. Frame, Chris Huntingford, Chris D. Jones, Jason A. Lowe, et al. "Warming Caused by Cumulative Carbon Emissions towards the Trillionth Tonne." *Nature* 458 (2009): 1163–1166.

Alliance of Small Island States. *Small Islands Call For Research On Survival Threshold*. 2012. http://aosis.org/small-islands-call-for-research-on-survival-threshold/ (accessed November 5, 2012).

Allison, I., N. L. Bindoff, R. A. Bindschadler, P. M. Cox, M. H. England, J. E. Francis, N. Gruber, A. M. Haywood, D. J. Karoly, G. Kaser, C. Le Quéré, T. M. Lenton, M. E. Mann, B. I. McNeil, N. de Noblet, A. J. Pitman, S. Rahmstorf, E. Rignot, H. J. Schellnhuber, S. H. Schneider, S. C. Sherwood, R. C. J. Somerville, K. Steffen, E. J. Steig, M. Visbeck, and A. J. Weaver. *The Copenhagen Diagnosis, 2009: Updating the World on the Latest Climate Science*. Sydney: The University of New South Wales Climate Change Research Centre, 2009.

Anderson, Elizabeth. *Value in Ethics and Economics*. Cambridge, MA: Harvard University Press, 1993.

Arrow, Kenneth J. "Discounting, Morality, and Gaming," in *Discounting and Intergenerational Equity*, edited by Paul R. Portney and John P. Weyant. Washington, DC: Resources for the Future, 1999.

Athanasiou, Tom and Paul Baer. *Dead Heat: Global Justice and Global Warming*. New York: Seven Stories Press, 2002.

Augustine. "The City of God," in *Nicene and Post-Nicene Fathers*, translated by Marcus Dods, edited by Philip Schaff and Kevin Knight. Vol. 2. Buffalo, NY: Christian Literature Publishing Co., 1887.

Baer, Paul. "Adaptation: Who Pays Whom?" in *Fairness in Adaptation to Climate Change*, edited by W. Neil Adger, Jouni Paavola, Saleemul Huq, and M. J. Mace. Cambridge, MA: MIT Press, 2006.

Baer, Paul, Tom Athanasiou, Sivan Kartha, and Eric Kemp-Benedict. *The Greenhouse Development Rights Framework: The Right to Development in a Climate Constrained World*. Berlin: Heinrich-Boll-Stiftung, Christian Aid, EcoEquity, and the Stockholm Environment Institute, 2008.

Banuri, Tariq and Niclas Hällström. "A Global Programme to Tackle Energy Access and Climate Change," *Development Dialogue* 61 (2012): 264–279.

Barry, Brian. "Sustainability and Intergenerational Justice," *Theoria* 45 (1997): 43–65.

Beckerman, Wilfred and Joanna Pasek. *Justice, Posterity, and the Environment*. Oxford: Oxford University Press, 2001.

Beitz, Charles. *The Idea of Human Rights*. Oxford: Oxford University Press, 2009.

Beitz, Charles. *Political Theory and International Relations*. Rev. ed. Princeton, NJ: Princeton University Press, 1999.

Bell, Derek. "Does Anthropogenic Climate Change Violate Human Rights?" *Critical Review of International Social and Political Philosophy* 14 (2011a): 99–124.

Bell, Derek. "Global Climate Justice, Historic Emissions, and Excusable Ignorance," *The Monist* 94 (2011b): 391–411.

Benn, Stanley I. "Personal Freedom and Environmental Ethics: The Moral Inequality of Species," in *Equality and Freedom: International and Comparative Jurisprudence*, edited by Gray L. Dorsey. Vol. 2. Dobbs Ferry: Oceana Publications, 1977.

Bentham, Jeremy. "Plan of Parliamentary Reform," in *The Works of Jeremy Bentham*, 1817. Online Library of Liberty, 2012. http://files.libertyfund.org/files/1922/0872.03_Bk.pdf (accessed October 29, 2012).

Bodansky, Daniel. "The Who What, and Wherefore of Geoengineering Governance," *Climatic Change* 121 (2013): 539–551.

Brecht, Bertolt. "A Worker Reads History." http://allpoetry.com/poem/8503711-A_Worker_Reads_History-by-Bertolt_Brecht (accessed December 26, 2013).

Brock, Gillian. *Global Justice: A Cosmopolitan Account*. Oxford: Oxford University Press, 2009.

Broome, John. *Climate Matters: Ethics in a Warming World*. New York and London: W.W. Norton and Co., 2012.

Broome, John. *Counting the Costs of Climate Change*. London: The White Horse Press, 1992.

Buchanan, Allen. "From Nuremburg to Kosovo: The Morality of Illegal International Legal Reform," in *Human Rights, Legitimacy, and the Use of Force*, by Allen Buchanan. Oxford: Oxford University Press, 2010.

Buchanan, Allen and Robert Keohane. "The Legitimacy of Global Governance Institutions," in *Human Rights, Legitimacy, and the Use of Force*, edited by Allen Buchanan. Oxford: Oxford University Press, 2010.

Budd, Malcom. *The Aesthetic Appreciation of Nature*. Oxford: Oxford University Press, 2002.

Butt, Daniel. *Rectifying International Injustice: Principles of Compensation and Restitution Between Nations*. Oxford: Oxford University Press, 2009.

Camacho, Alejandro E. "Assisted Migration: Redefining Nature and Natural Resource Law under Climate Change," *Yale Journal on Regulation* 27 (2010): 171–255.

Caney, Simon. "Just Emissions," *Philosophy and Public Affairs* 40 (2012): 255–300.

Caney, Simon. "Climate Change and the Duties of the Advantaged," *Critical Review of International Social and Political Philosophy* 13 (2010a): 203–228.

Caney, Simon. "Climate Change, Human Rights, and Moral Thresholds," in *Human Rights and Climate Change*, edited by Stephen Humphreys. Cambridge: Cambridge University Press, 2010b.

Caney, Simon. "Human Rights, Responsibilities, and Climate Change," in *Global Basic Rights*, edited by Charles R. Beitz and Robert E. Goodin. Oxford: Oxford University Press, 2009.

Caney, Simon. "Environmental Degradation, Reparations, and the Moral Significance of History," *Journal of Social Philosophy* 37 (2006a): 464–482.

Caney, Simon. *Justice Beyond Borders: A Global Political Theory*. Oxford: Oxford University Press, 2006b.

Caney, Simon. "Cosmopolitan Justice, Responsibility, and Global Climate Change," *Leiden Journal of International Law* 18 (2005): 747–775.

Carbon Tracker. "Unburnable Carbon 2013: Wasted capital and stranded assets." http://www.carbontracker.org/wp-content/uploads/downloads/2011/07/Unburnable-Carbon-Full-rev2.pdf (accessed December 18, 2013).

Carlson, Alan. "Nature and Positive Aesthetics," *Environmental Ethics* 6 (1984): 5–34.

Carter, Alan. "Biodiversity and All the Jazz," *Philosophy and Phenomenological Research* 80 (2010): 58–75.

Chen, I-Ching, Jane K. Hill, Ralf Ohlemüller, David B. Roy, and Chris D. Thomas. "Rapid Range Shifts of Species Associated with High Levels of Climate Warming," *Science* 333 (2011): 1024–1026.

Cicerone, Ralph J. "Geoengineering: Encouraging Research and Overseeing Implementation," *Climatic Change* 77 (2006): 221–226.

Cline, Willam. *Global Warming and Agriculture*. Washington, DC: Peterson Institute, 2007.

Conway, Eric M. and Naomi Oreskes. *Merchants of Doubt*. New York: Bloomsbury, 2010.

Cranor, Carl F. *Regulating Toxic Substances: A Philosophy of the Science and the Law*. New York: Oxford University Press, 1993.

Crutzen, Paul J. "Albedo Enhancement by Stratospheric Sulfur Injections: A Contribution to Resolve a Policy Dilemma?" *Climatic Change* 77 (2006): 211–220.

Daly, Herman. *Steady State Economics*, 2nd ed. Washington, DC: Island Press, 1991.

Darwall, Stephen. *The Second Person Standpoint: Morality, Respect and Accountability*. Cambridge, MA: Harvard University Press, 2006.

Dasgupta, Partha. "Discounting Climate Change," *Journal of Risk and Uncertainty* 37 (2008): 141–169.

Dasgupta, Partha, Scott Barrett, and Karl-Göran Mäler. "Intergenerational Equity Social Discount Rates, and Global Warming," in *Discounting and Intergenerational Equity*, edited by Paul R. Portney and John P. Weyant. Washington, DC: Resources for the Future, 1999.

Dessai, Suraje, W. N. Adger, Mike Hulme, Jonathan Koehler, Jonathan Turnpenny, and Rachel Warren. "Defining and Experiencing Dangerous Climate Change," *Climate Change* 64 (2004): 11–25.

Dessai, Suraje, Mike Hulme, Robert Lempert, and Roger, Jr., Pielke. "Climate Prediction: A Limit to Adaptation," in *Adapting to Climate Change: Thresholds, Values, and Governance*, edited by W. Neil Adger, Irene Lorenzoni, and Karen L. O'Brien. Cambridge: Cambridge University Press, 2009.

Dewey, John. *Reconstruction in Philosophy*, enlarged ed. Boston: Beacon, 1957.

Doran, Peter T. and Maggie Kendall Zimmerman. "Examining the Scientific Consensus on Climate Change," *EOS, Transactions, American Geophysical Union* 90 (2009): 22–23.

Dworkin, Ronald. *Life's Dominion: An Argument about Abortion, Euthanasia, and Individual Freedom*. New York: Vintage Books, 1994.

Dworkin, Ronald. *Taking Rights Seriously*. Cambridge, MA: Harvard University Press, 1978.

Elliot, Robert. *Faking Nature: The Ethics of Environmental Restoration*. London and New York: Routledge, 1997.

Emerson, Ralph Waldo. *Nature*. 1836; Oregon State University, 2013. http://oregon state.edu/instruct/phl302/texts/emerson/nature-emerson-a.html#Introduction (accessed October 29, 2012).

Emission Data for Global Atmospheric Research (EDGAR). CO_2 *time series 1990–2011 per region/country*. 2011. http://edgar.jrc.ec.europa.eu/overview.php?v=CO2ts 1990-2011 (accessed September 10, 2012).

Fellner, William. *Probability and Profit*. Homewood: Richard D. Irwin, Inc., 1965.

Fleurbaey, Marc and Bertil Tungodden. "The Tyranny of Non-Aggregation versus the Tyranny of Aggregation in Social Choices: A Real Dilemma," *Economic Theory* 44 (2010): 399–414.

Forst, Rainer. *The Right to Justification*. Translated by Jeffrey Flynn. New York: Columbia University Press, 2011.

Forst, Rainer. "The Justification of Human Rights and the Basic right to Justification: A Reflexive Approach," *Ethics* 120 (2010): 711–740.

Gardiner, Stephen M. *A Perfect Moral Storm: The Ethical Tragedy of Climate Change*. Oxford: Oxford University Press, 2011.

Gardiner, Stephen M. "Rawls and Climate Change: Does Rawlsian Political Philosophy Pass the Global Test?" *Critical Review of International Social and Political Philosophy* 14 (2011): 125–151.

Gardiner, Stephen M. "Ethics and Climate Change: An Introduction," *WIREs* 54 (2010a): 54–66.

Gardiner, Stephen M. "Is 'Arming the Future' with Geoengineering Really the Lesser Evil? Some Doubts About the Ethics of Intentionally Manipulating the Climate System," in *Climate Ethics: Essential Readings*, edited by Stephen Gardiner, Simon Caney, Dale Jamieson, and Henry Shue. Oxford: Oxford University Press, 2010b.

Gardiner, Stephen M. "A Core Precautionary Principle," *Journal of Political Philosophy* 14 (2006): 33–60.

Gardiner, Stephen M., Simon Caney, Dale Jamieson, and Henry Shue, eds. *Climate Ethics: Essential Readings*. Oxford: Oxford University Press, 2010.

Garvey, James. *The Ethics of Climate Change: Right and Wrong in a Warming World*. London: Continuum, 2008.

Gollier, Christian and Nicolas Treich. "Decision-Making under Scientific Uncertainty: The Economics of the Precautionary Principle," *Journal of Risk and Uncertainty* 27 (2003): 77–103.

Goodin, Robert E. "Toward an International Rule of Law: Distinguishing International Law Breakers from Would-Be Law-Makers," *Journal of Ethics* 9 (2005): 225–246.

Goodin, Robert E. *Green Political Theory*. London: Polity Press, 1992.

Griffin, James. *On Human Rights*. Oxford: Oxford University Press, 2008.

Grosseries, Axel. "Historical Emissions and Free-Riding," *Ethical Perspectives* 11 (2004): 36–60.

Grotius, Hugo. *The Rights of War and Peace*, BK. III, edited by Richard Tuck. Indianapolis, IN: Liberty Fund, 2005.

Habermas, Jürgen. "Discourse Ethics: Notes on a Program of Philosophical Justification," in *Moral Consciousness and Communicative Action*, translated by Christian Lenhardt and Shierry Weber Nicholsen. Cambridge: Massachusetts Institute of Technology Press, 1991.

Hansen, James. "Huge Sea Level Rises are Coming – Unless We Act Now," *New Scientist* 2614, (2007). http://www.newscientist.com/article/mg19526141.600-huge-sea-level-rises-are-coming-unless-we-act-now.html?full=true (accessed September 6, 2012).

Harris, Paul. *World Ethics and Climate Change: From International Justice to Global Justice*. Edinburgh: University of Edinburgh Press, 2010.

Hart, H. L. A. and A. M. Honoré. *Causation in the Law*. Oxford: Oxford University Press, 1959.

Hassoun, Nicole. "The Anthropocentric Advantage? Environmental Ethics and Climate Change Policy," *Critical Review of International Social and Political Philosophy* 14 (2011): 235–257.

Hayward, Tim. "Human Rights versus Emissions Rights," *Ethics and International Affairs* 21 (2007): 431–450.

Heal, G. M. "Intertemporal Welfare Economics and the Environment," in *Handbook of Environmental Economics*, edited by Karl-Göran Mäler and Jeffrey Vincent, Vol. 3. Amsterdam: Elsevier, 2005.

Hill, Thomas E., Jr. "Ideals of Human Excellence and Preserving Natural Environments," *Environmental Ethics* 5 (1983): 211–224.

Hobbes, Thomas. *The Leviathan*. Indianapolis: Hackett, 1994.

Hönisch, Bärbel, Stephen Barker, Gavin L. Foster, Samantha J. Gibbs, Sarah E. Greene, Wolfgang Kiessling, Lee Kump, Thomas M. Marchitto Jr., Rowan C. Martindale, Ryan Moyer, Carles Pelejero, Andy Ridgwell, Justin Ries, Dana L. Royer, Daniela N. Schmidt, Appy Sluijs, Ellen Thomas, Branwen Williams, James C. Zachos, Richard Zeebe, and Patrizia Ziveri. "The Geological Record of Ocean Acidification," *Science* 335 (2012): 1058–1063.

Honoré, Tony. *Responsibility and Fault*. Oxford and Portland: HartPublishing, 1999.

Howarth, Richard B. "An Overlapping Generations Model of Climate-Economy Interactions," *The Scandinavian Journal of Economics* 100 (1998): 575–591.

Hulme, Mike. *Why We Disagree about Climate Change*. Cambridge: Cambridge University Press, 2009.

Intergovernmental Panel on Climate Change. *Climate Change: A Synthesis Report: Summary for Policy Makers*. 2007a. http://www.ipcc.ch/pdf/assessment-report/ar4/syr/ar4_syr_spm.pdf (accessed January 11, 2011).

Intergovernmental Panel on Climate Change. Climate Change 2007: The Physical Science Basis. Contribution of Working Group I to the Fourth Assessment Report of the Intergovernmental Panel on Climate Change. 2007b. http://www.ipcc.ch/pdf/assessment-report/ar4/wg1/ar4-wg1-chapter10.pdf (accessed February 20, 2009).

International Covenant on Economic, Social and Cultural Rights. http://www2.ohchr
.org/english/law/cescr.htm (accessed December 26, 2012).

International Energy Agency. *Energy Poverty: How to Make Modern Energy Access Universal.* 2010. http://www.iea.org/publications/worldenergyoutlook/resources/ energydevelopment/universalenergyaccess/ (accessed November 22, 2012).

Jamieson, Dale. *Reason in a Dark Time: Why the Struggle Against Climate Change Failed and What It Means for Our Future.* Oxford: Oxford University Press, 2014.

Jamieson, Dale. "Some Whats, Whys, and Worries of Geoengineering," *Climatic Change* 121(2013): 527–537.

Jamieson, Dale. "Climate Change, Responsibility, and Justice," *Science and Engineering Ethics* 16 (2010): 431–445.

Jamieson, Dale. "Ethics and Intentional Climate Change," *Climatic Change* 33 (1996): 323–336.

Jamieson, Dale. "Ethics, Public Policy, and Global Warming," *Science, Technology, and Human Values* 17 (1992): 139–153.

Jamieson, Dale. "Managing the Future: Public Policy, Scientific Uncertainty, and Global Warming," in *Upstream/Downstream: Issues in Environmental Ethics*, edited by Donald Scherrer. Philadelphia, PA: Temple University Press, 1990.

Kant, Immanuel. *Groundwork of the Metaphysics of Morals*, translated and edited by Mary Gregor. Cambridge and New York: Cambridge University Press, 1998.

Kant, Immanuel. *Grounding for the Metaphysics of Morals: On a Supposed Right to Lie Because of Philanthropic Concerns*, translated by James W. Ellington, 3rd ed. Indianapolis: Hackett Publishing, 1993.

Kant, Immanuel. "The Metaphysics of Morals," in *Immanuel Kant, Ethical Philosophy*, translated by James W. Ellington. Indianapolis and Cambridge: Hackett Publishing, 1983.

Kant, Immanuel. "Idea for a Universal History with a Cosmopolitan Purpose," in Kant's *Political Writings*, edited by Hans Reiss, and translated by H. B. Nisbet. Cambridge: Cambridge University Press, 1970.

Karnein, Anja. "Putting Fairness in Its Place: Why There Is a Duty to Take Up the Slack," Lecture presented at the University of Edinburgh, Scotland, June 2012.

Kennedy, Martin, David Mrofka, and Chris von der Borch. "Snowball Earth Termination by Destabilization of Equatorial Permafrost Methaneclathrate," *Nature* 453 (2008): 642–645.

Keohane, Robert O. and David G. Victor. "The Regime Complex for Climate Change," *Perspective on Politics* 9 (2011): 7–23.

Keohane, Robert. *After Hegemony: Cooperation and Discord in the World Political Economy.* Princeton, NJ: Princeton University Press, 1984.

King, Jr., Martin Luther. "I Have a Dream." http://www.ushistory.org/documents/ i-have-a-dream.htm (accessed December 18, 2013).

Knight, Frank H. *Risk, Uncertainty, and Profit.* Boston: Hart, Schaffner and Marx; Houghton Mifflin Co., 1921.

Kolbert, Elizabeth. "Enter the Anthropocene – The Age of Man," *National Geographic*, March 2011. http://ngm.nationalgeographic.com/2011/03/age-of-man/kolbert-text.

Koopmans, Tjalling C. "Stationary Ordinal Utility and Impatience," *Ecomentrica* 28 (1960): 287–309.

Latham, John, Keith Bower, Tom Choularton, Hugh Coe, Paul Connolly, Gary Cooper, Tim Craft, Jack Foster, Alan Gadian, Lee Galbraith, Hector Iacovides, David Johnston, Brian Launder, Brian Leslie, John Meyer, Armand Neukermans, Bob Ormond, Ben Parkes, Phillip Rasch, John Rush, Stephen Salter, Tom Stevenson, Hailong Wang, Qin Wang, and Rob Wood. "Marine Cloud Brightening," *Philosophical Transactions of the Royal Society* A 370 (2012): 4217–4262.

Leopold, Aldo. "The Land Ethic," in *The Sand County Almanac*, edited by Aldo Leopold. 1948. http://home.btconnect.com/tipiglen/landethic.html (accessed March 30, 2012).

Lin, Bing, Lin Chambers, Yongxiang Hu, Bruce Wielicki, and Kuan-Man Xu. "The Iris Hypothesis: A Negative or Positive Cloud Feedback?" *Journal of Climate* 15 (2002): 3–7.

Lindzen, Richard S. "Is the Global Warming Alarm Founded on Fact?" in *Global Warming: Looking Beyond Kyoto*, edited by Ernesto Zedillo. Washington, DC: The Brookings Institution Press, 2008.

Llavador, Humberto, John E. Roemer, and Joaquim Silvestre. "North-South Convergence and the Allocation of CO_2 Emissions." Working paper, Barcelona Graduate School of Economics, 2010.

Locke, John. "A Letter Concerning Toleration," in *John Locke, Political Writings*, edited by David Wootton. Indianapolis: Hackett Publishing, 1993.

Lodge, David M. and Kristin Shrader-Frechette. "Nonindigenous Species: Ecological Explanation, Environmental Ethics, and Public Policy," *Conservation Biology* 17 (2003): 31–37.

Lomborg, Bjørn. *Cool It: The Skeptical Environmentalist's Guide to Global Warming*. New York: Alfred A. Knopf, 2008.

Manson, Neil A. "Formulating the Precautionary Principle," *Environmental Ethics* 24 (2002): 192–202.

Marlene-Russow, Lilly. "Why Do Species Matter?" *Environmental Ethics* 3 (1981): 101–112.

Marx, Karl. "On The Jewish Question," in *Karl Marx: Selected Writings*, edited by David McLellan. Oxford: Oxford University Press, 1977.

Matthews, H. Damon, Nathan P. Gillett, Peter A. Stott, and Kirsten Zickfeld. "The Proportionality of Global Warming to Cumulative Carbon Emissions," *Nature* 459 (2009): 829–832.

Matthews, H. Damon, Susan Solomon, and Raymond Pierrehumbert. "Cumulative Carbon as a Policy Framework for Achieving Climate Stabilization," *Philosophical Transactions of the Royal Society* A 370 (2012): 4365–4379.

McGranahan, Gordon, Deborah Balk, and Bridget Anderson. "The Rising Tide: Assessing the Risks of Climate Change and Human Settlements in Low Elevation Coastal Zone." *Environment and Urbanization* 19 (2007): 17–37.

McLachlan, Jason S., Jessica J. Hellmann, and Mark W. Schwartz. "A Framework for Debate of Assisted Migration in an Era of Climate Change," *Conservation Biology* 21 (2007): 297–302.

Meehl, Gerald A., W. D. Collins, P. Friedlingstein, A. T. Gaye, J. M. Gregory, A. Kitoh, R. Knutti, J. M. Murphy, A. Noda, S. C. B. Raper, T. F. Stocker, I. G. Watterson,

A. J. Weaver, and Z. -C. Zhao. "Global Climate Projections," in *Climate Change 2007: The Physical Science Basis. Contribution of Working Group I to the Fourth Assessment Report of the Intergovernmental Panel on Climate Change*, edited by S. Solomon, et al. Cambridgeand New York: Cambridge University Press, 2007.

Meehl, Gerald A., Julie M. Arblaster, Lawrence E. Buja, William D. Collins, Aixue Hu, Warren G. Strand, Haiyan Teng, and Warren M. Washington. "How Much More Global Warming and Sea Level Rise?" *Science* 307 (2005): 1769–1772.

Meinshausen, Malte, Myles R. Allen, David J. Frame, Katja Frieler, William Hare, Reto Knutti, Nicolai Meinshausen, and Sarah C. B. Raper. "Greenhouse-Gas Emission Targets for Limiting Global Warming to 2°C," *Nature* 458 (2009): 1158–1162.

Meyer, Lukas H. and Dominic Roser. "Distributive Justice and Climate Change: The Allocation of Emission Rights," *Analyse & Kritik* 2 (2006): 223–249.

Mill, John Stuart. "Utilitarianism," in *The Collected Works of John Stuart Mill*, edited by J. M. Robson. Toronto: The University of Toronto Press, 2006.

Mill, John Stuart. *Principles of Political Economy: With Some of Their Applications to Social Philosophy*. London: J. W. Parker, 1848.

Miller, David. "Global Justice and Climate Change: How Should Responsibilities Be Distributed?" The Tanner Lectures on Human Values, Tsinghua University, Beijing, March 24–25, 2008.

Miller, David. *National Responsibility and Global Justice*. Oxford: Oxford University Press, 2007.

Minteer, Ben A. and James P. Collins. "Move It or Lose It? The Ecological Ethics of Relocating Species under Climate Change," *Ecological Applications* 20 (2010): 1801–1804.

Moellendorf, Darrel. "A Right to Sustainable Development," *The Monist* 94 (2011a): 433–452.

Moellendorf, Darrel. "A Normative Account of Dangerous Climate Change," *Climatic Change* 108 (2011b): 57–72.

Moellendorf, Darrel. *Global Inequality Matters*. Basingstoke: Palgrave Macmillan, 2009a.

Moellendorf, Darrel. "Treaty Norms and Climate Change Mitigation," *Ethics and International Affairs* 23 (2009b): 247–265.

Moellendorf, Darrel. "Justice and the Assignment of the Intergeneartional Costs of Climate Change," *Journal of Social Philosophy*, 40 (2009c): 204–224.

Moellendorf, Darrel. *Cosmopolitan Justice*. Boulder: Westview Press, 2002.

NASA, Earth Observatory. "Global Warming." June 3, 2010a. http://earthobservatory .nasa.gov/Features/GlobalWarming/page2.php (accessed September 5, 2012).

NASA, Earth Observatory. "How is Today's Warming Different from the Past?" June 3, 2010b. http://earthobservatory.nasa.gov/Features/GlobalWarming/page3.php (accessed September 6, 2012).

National Oceanography Centre, Southampton. *Warming Ocean Contributes to Global Warming*. 2009. http://www.noc.soton.ac.uk/nocs/news.php?action=display_ news&idx=628 (accessed August 31, 2012).

National Snow and Ice Data Center. Icelights: Your Burning Questions about Ice & Climate: What Does Seeping Methane Mean for the Thawing Arctic? 2012. http://nsidc .org/icelights/2012/07/03/what-does-seeping-methane-mean-for-the-thawing-arctic/ (accessed August 31, 2012).

Neumayer, Eric. "In Defence of Historical Accountability for Greenhouse Emissions," *Ecological Economics* 33 (2000): 185–192.

New, Mark, Kevin Anderson, Diana Liverman, and Heike Schroder. "Four Degrees and Beyond: The potential for Global Temperature Increase of Four Degrees and Its Implications," *Philosophical Transactions of the Royal Society* A 369 (2011): 6–19.

Nickel, James W. *Making Sense of Human Rights*, 2nd ed. Malden: Blackwell Publishing, 2007.

Nietzsche, Friedrich. *On the Genealogy of Morality*, translated by Carol Diethe, edited by Kieth Ansell-Pearson. Cambridge: Cambridge University Press, 2006.

Nietzsche, Friedrich. *The Gay Science*, translated by Josefine Nauckhoff, edited by Bernard Williams. Cambridge: Cambridge University Press, 2001.

Nordhaus, William. *A Question of Balance: Weighing the Options on Global Warming Policies*. New Haven and London: Yale University Press, 2008.

Northcott, Michael S. *A Moral Climate: The Ethics of Global Warming*. New York: Orbis Books, 2009.

Norton, Bryan. "Ecology and Opportunity: Intergenerational Equity and Sustainable Options," in *Fairness and Futurity*, edited by Andrew Dobson. Oxford: Oxford University Press, 1999.

Norton, Bryan G. "On the Inherent Danger of Undervaluing Species," in *The Preservation of Species*, edited by Bryan G. Norton. Princeton, NJ: Princeton University Press, 1986.

Nozick, Robert. *Anarchy, State and Utopia*. New York: Basic Books, 1977.

Nozick, Robert. "Newcomb's Problem and Two Principles of Choice," in *Essays in Honor of Karl G. Hempel*, edited by Nicholas Rescher. Dordrecht, Netherlands: Kluwer, 1969.

O'Neill, Brian C. and Michael Oppenheimer. "Dangerous Climate Impacts and the Kyoto Protocol," *Science* 296 (2002): 1971–1972.

Osler, Fen and Judith Reppy, eds. *Earthly Goods*. Ithaca, NY: Cornell University Press, 1996.

Pachauri, Rajendra. "Avoiding Dangerous Climate Change," in *Avoiding Dangerous Climate Change*, edited by Wolfgang Cramer, Nebojsa Nakicenovic, Hans Joachim Schellnhuber, Tom Wigley, and Gary Yohe. Cambridge: Cambridge University Press, 2006.

Page, Edward. *Climate Change, Justice and Future Generations*. Cheltennam, UK: Edward Elgar Publishing, 2006.

Page, Edward A. "Give It Up for Climate Change: A Defence of the Beneficiary Pays Principle," *International Theory* 4 (2012): 313–317.

Plato. *Gorgias*, translated by Walter Hamilton. London: Penguin Classics, 2004.

Pogge, Thomas. *World Poverty and Human Rights*, 2nd ed. London: Polity, 2008.

Pogge, Thomas. *Realizing Rawls*. Ithaca: Cornell University Press, 1989.

Poore, Richard Z., Christopher Tracey, and Richard S. Williams, Jr. "Sea Level and Climate," United States Geological Survey Fact Sheet 020–00. http://pubs.usgs.gov/fs/fs2-00/ (accessed July 4, 2013).

Posner, Eric A. and David Weisbach. *Climate Change Justice*. Princeton, NJ: Princeton University Press, 2010.

Posner, Richard A. *Catastrophe: Risk and Response*. Oxford: Oxford University Press, 2004.

Pufendorf, Samuel. *On the Duty of Man and Citizen*, edited by James Tully. Cambridge: Cambridge University Press, 1991.

Rahmstorf, Stefan, Anny Cazenave, John A. Church, James E. Hansen, Ralph F. Keeling, David E. Parker, and Richard C. J. Somerville. "Recent Climate Observations Compared to Predictions," *Science* 316 (2007): 709.

Randall, Alan. "Human Preferences, Economics, and Preservation," in *The Preservation of Species*, edited by Bryan G. Norton. Princeton, NJ: Princeton University Press, 1986.

Ramsey, F. P. "A Mathematical Theory of Saving," *The Economic Journal* 38 (1928): 543–559.

Ramsey, Frank P. "Truth and Probability," in *The Foundations of Mathematics and Other Logical Essays*, edited by R. B. Braithwaite. London and New York: Routledge, 1931.

Rawls, John. *The Law of Peoples*. Cambridge, MA: Harvard University Press, 1999a.

Rawls, John. *A Theory of Justice*, rev. ed. Cambridge, MA: Harvard University Press, 1999b.

Rayner, Steve, Clare Heyward, Tim Kruger, Nick Pidgeon, Catherin Redgwell, and Julian Savulescu, "The Oxford Principles," *Climatic Change* 121 (2013): 499–512.

Raz, Joseph. *The Morality of Freedom*. Oxford: Oxford University Press, 1986.

Richardson, Katherine, Joseph Alcamo, Terry Barker, Hans Joachim Schellnhuber, Daniel M. Kammen, Rik Leemans, Diana Liverman, Mohan Munasinghe, Balgis Osman-Elasha, Will Stefen, Nicholas Stern, and Ole Waever. *Synthesis Report*, proceedings from the conference "Climate Change, Global Risks, Challenges, and Decisions" at the University of Copenhagen, March 10–12, 2009. http://climatecongress.ku.dk/pdf/synthesisreport (accessed March 4, 2010).

Ridgwell, Andy, Chris Freeman, and Richard Lampitt. "Geoengineering: Taking Control of Our Planet's Climate?" *Philosophical Transactions of the Royal Society A* 370 (2012): 4163–4165.

Roberts, J. Timmons and Bradley C. Parks. *A Climate of Injustice: Global Inequality, North-South Politics, and Climate Policy*. Cambridge, MA: Massachusetts Institute of Technology Press, 2007.

Rolston, Holmes, III. "Value in Nature and the Nature of Value," in *Philosophy and the Natural Environment*, edited by Robin Attfield and Andrew Belsey. Cambridge: Cambridge University Press, 1994.

Routley, Richard and Val Routley. "Against the Inevitability of Human Chauvinism," in *Ethics and the Problems of the 21st Century*, edited by K. E. Goodpaster and K. M. Sayre. South Bend, IN: Notre Dame University Press, 1979.

Sachs, Wolfang, ed. *The Development Dictionary*, 2nd ed. London: Zed Books, 2010.

Sagoff, Mark. *The Economy of the Earth: Philosophy, Law, and the Environment*, 2nd ed. Cambridge: Cambridge University Press, 2008.

Sandler, Ronald. "The Value of Species and the Ethical Foundations of Assisted Colonization," *Conservation Biology* 24 (2010): 424–431.

Scanlon, T. M. *What We Owe to Each Other*. Cambridge, MA: Harvard University Press, 1998.

Schaffer, Axel and Darrel Moellendorf. "Beyond Discounted Utilitarianism – Just Distribution of Climate Costs," *Karlsruhe Beiträge zur Wirtschaftspolitischen Forschung* 33 (2013) forthcoming.

Schellnhuber, Hans Joachim, Wolfgang Cramer, Nebojsa Nakicenovic, Tom Wigley, and Gary Yohe, eds. *Avoiding Dangerous Climate Change*. Cambridge: Cambridge University Press, 2006.

Schneider, Stephen H. "What is 'Dangerous' Climate Change, *Nature* 411 (2001): 17–19.

Schneider, Stephen H. "Geoengineering: Could – or Should – We Do It?" *Climatic Change* 33 (1996): 291–302.

Schneider, Stephen H. and Kristin Kuntz-Duriseti. "Uncertainty and Climate Change Policy," in *Climate Change Policy: A Survey*, edited by John O. Niles, Armin Rosencranz, and Stephen H. Schneider. Washington, DC: Island Press, 2002.

Schwartz, Mark W., Justin O. Borevitz, Jean Brennan, Alex E. Camacho, Gerardo Ceballos, Jamie R. Clark, Holly Doremus, Regan Early, Julie R. Etterson, Dwight Fielder, Jaqueline L. Gill, Patrick Gonzalez, Nancy Green, Lee Hannah, Jessica J. Hellmann, Dale W. Jamieson, Debra Javeline, Jason M. McLachlan, Ben A. Minteer, Jay Odenbaugh, Stephen Polasky, David M. Richardson, Terry L. Root, Hugh D. Safford, Osvaldo Sala, Dov F. Sax, Stephen H. Schneider, Andrew R. Thompson, John W. Williams, Mark Vellend, Pati Vitt, and Sandra Zellmer. "Managed Relocation: Integrating the Scientific, Regulatory, and Ethical Challenges," *BioScience* 62 (2012): 732–743.

Sen, Amartya. *Development as Freedom*. New York: Anchor, 2000.

Sen, Amartya. "Rights and Agency," *Philosophy and Public Affairs* 11 (1982): 3–39.

Shoo, Luke P., Ary A. Hoffmann, Stephen Garnett, Robert L. Pressey, Yvette M. Williams, Martin Taylor, Lorena Falconi, Colin J. Yates, John K. Scott, Diogo Alagador, and Stephen E. Williams. "Making Decisions to Conserve Species Under Climate Change." *Climatic Change* 119 (2013): 239–246.

Shue, Henry. *Climate Justice: Vulnerability and Protection*. Oxford: Oxford University Press, 2014.

Shue, Henry. "Climate Hope: Implementing the Exit Strategy," *Chicago Journal of International Law* 13 (2012): 394–395.

Shue, Henry. "Deadly Delays, Saving Opportunities: Creating a More Dangerous World?" in *Climate Ethics: Essential Readings*, edited by Stephen Gardiner, Simon Caney, Dale Jamieson, and Henry Shue. Oxford: Oxford University Press, 2010.

Shue, Henry. "Responsibility to Future Generations and the Technological Transition," in *Perspectives on Climate Change: Science, Economics, Politics, Ethics*, edited by Walter Sinnot-Armstrong and Richard B. Howarth. Amsterdam and San Diego: Elsevier, 2005.

Shue, Henry. "Bequeathing Hazards: Security Rights and Property Rights of Future Humans," in *Limits to Markets: Equity and the Global Environment*, edited by Mohamed Dore and Timothy Mount. Malden, MA: Blackwell Publishers, 1998.

Shue, Henry. "Eroding Sovereignty: The Advance of Principle," in *The Morality of Nationalism*, edited by Robert McKim and Jeff McMahan. Oxford: Oxford University Press, 1997.

Shue, Henry. "Avoidable Necessity: Global Warming, International Fairness and Alternative Energy," in *Theory and Practice*, edited by Ian Shapiro and Judith Wagner DeCew. New York: New York University Press, 1995.

Shue, Henry. "After You: May Action by the Rich Be Contingent Upon Action by the Poor?" *Indiana Journal of Global Legal Studies* 1 (1994): 352–353.

Shue, Henry. "Subsistence Emissions and Luxury Emissions," *Law & Policy* 15 (1993): 39–59.

Shue, Henry. "The Unavoidability of Justice," in *The International Politics of the Environment: Actors, Interests, and Institutions*, edited by Andrew Hurrel and Benedict Kingsbury. Oxford: Oxford University Press, 1992.

Singer, Peter. "Famine, Affluence, and Morality," *Philosophy and Public Affairs* 1 (1972): 229–243.

Smith, Mark Stafford, Clive Hamilton, Alex Harvey, andLisa Horrocks. "Rethinking Adaptation for a 4°C World," *Philosophical Transactions of the Royal Society A* 369 (2011): 196–216.

Sober, Elliott. "Philosophical Problems for Environmentalism," in *The Preservation of Species*, edited by Bryan G. Norton. Princeton, NJ: Princeton University Press, 1986.

Specter, Michael. "The Climate Fixers," *New Yorker*, May 14, 2012.

Stern, Nicholas. "Stern Review on the Economics of Climate Change." http://webarchive.nationalarchives.gov.uk/20130129110402/http://www.hm-treasury.gov.uk/sternreview_index.htm (accessed October 5, 2012a).

Stern, Nicholas. "Stern Review on the Economics of Climate Change: Summary of the Conclusions." http://webarchive.nationalarchives.gov.uk/20130129110402/http://www.hm-treasury.gov.uk/d/Summary_of_Conclusions.pdf (accessed September 6, 2012).

Stern, Nicholas. *The Global Deal: Climate Change and the Creation of a New Era of Progress and Prosperity.* New York: Public Affairs Books, 2009.

Sunstein, Cass R. *Laws of Fear.* Cambridge: Cambridge University Press, 2005.

Taylor, Paul W. *Respect for Nature.* Princeton, NJ: Princeton University Press, 2001.

Tedeschi, Bob. "Safeguarding Against Loan Discrimination," *New York Times*, January 25, 2009. http://www.nytimes.com/2009/01/25/realestate/25mort.html?scp=5&sq=home_lending_racial_discrimination&st=cse (accessed May 7, 2009).

Trillionth Tonne. "Explaining the Need to Limit Cumulative Emissions of Carbon Dioxide." http://trillionthtonne.org/ (accessed December 18, 2013).

Union of Concerned Scientists. *Each Country's Share of CO_2 Emissions.* 2012. http://www.ucsusa.org/global_warming/science_and_impacts/science/each-countrys-share-of-co2.html (accessed November 5, 2012).

United Nations. *Universal Declaration of Human Rights.* Geneva: UN, 1948. http://www.un.org/en/documents/udhr/index.shtml (accessed September 13, 2012).

United Nations. *Report of the World Commission on Environment and Development.* Geneva: UN, 1987. http://www.un.org/documents/ga/res/42/ares42-187.htm (accessed November 1, 2012).

United Nations Children's Fund (UNICEF). *State of the World's Children 2012 Children in an Urban World, Executive Summary.* New York: UNICEF, 2012. http://www.unicef.org/publications/files/SOWC_2012-Executive_Summary_EN_13 Mar2012.pdf (accessed November 1, 2012).

United Nations Educational, Scientific, and Cultural Organization. *A 30-Year Struggle: The Sustained Effort to Give Force of Law to the Universal Declaration on Human Rights.* Paris: UNESCO, 1977. http://unesdoc.unesco.org/images/0007/000748/074816eo.pdf#48063 (accessed December 27, 2012).

United Nations Environment Programme. *The Emissions Gap Report.* New York: UNEP, 2010. http://www.unep.org/publications/ebooks/emissionsgapreport/pdfs/GAP_REPORT_SUNDAY_SINGLES_LOWRES.pdf (accessed November 22, 2012).

United Nations Environment Programme. *Rio Declaration on Environment and Development*. Rio de Janeiro: UNEP, 1992. http://www.unep.org/Documents.Multi lingual/Default.asp?documentid=78&articleid=1163 (accessed November 1, 2012).

United Nations Framework Convention on Climate Change. *Report of the Conference of the Parties on Its Seventeenth Session*. Durban: 2011. http://unfccc.int/resource/docs/2011/cop17/eng/09a01.pdf#page=2 (accessed November 22, 2012).

United Nations Framework Convention on Climate Change. *Report of the Conference of the Parties of Its Sixteenth Session*. Cancún: 2010. http://unfccc.int/resource/docs/2010/cop16/eng/07a01.pdf#page=2 (accessed November 22, 2012).

United Nations Framework Convention on Climate Change. *SBSTA Technical Briefing: Historical Responsibility*. Bonn: UNFCC, 2009. http://unfccc.int/files/mee tings/ad_hoc_working_groups/lca/application/pdf/1_shue_rev.pdf (accessed June 27, 2013).

United Nations Framework Convention on Climate Change. *Investment and Financial Flows to Address Climate Change, Executive Summary*. 2007. http://unfccc. int/files/cooperation_and_support/financial_mechanism/application/pdf/executive_ summary.pdf (accessed December 19, 2012).

United Nations Framework Convention on Climate Change. *Ad Hoc Group on the Berlin Mandate, Seventh Session, Bonn, 1997, Implementation of the Berlin Mandate, Addendum, Note by Secretariat*. Bonn: 1997. http://unfccc.int/resource/docs/1997/ agbm/misc01a03.pdf (accessed December 12, 2012).

United Nations Framework Convention on Climate Change. Geneva: 1992. http:// unfccc.int/essential_background/convention/background/items/1349.php (accessed January 15, 2011).

United Nations Human Development Programme. *Human Development Report 2009, Overcoming Barriers: Human Mobility and Development*. New York: UNEP, 2009. http://hdr.undp.org/en/media/HDR_2009_EN_Complete.pdf (accessed November 22, 2012).

United Nations Human Development Programme. *Human Development Programme 2007/2008 Fighting Climate Change: Human Solidarity in a Divided World*. New York: UNDP, 2007. http://hdr.undp.org/en/media/HDR_20072008_EN_Complete .pdf (accessed April 29, 2013).

United States Congress. *Endangered Species Act of 1973*. Washington, DC, 1973.

United States Congressional Budget Office. *The Estimated Costs to Households from the Cap-and-Trade Provisions of H.R. 2454*. Washington, DC, 2009a. http://www.cbo.gov/sites/default/files/cbofiles/ftpdocs/103xx/doc10327/06-19-capand tradecosts.pdf (accessed July 2, 2013).

United States Congressional Budget Office. *Economic and Budget Issue Brief*. Washington, DC, 2009b. http://www.cbo.gov/ftpdocs/104xx/doc10458/11-23-Greenhouse GasEmissions_Brief.pdf (accessed November 1, 2012).

United States Department of Commerce, NOAA. "Trends in Carbon Dioxide." http://www.esrl.noaa.gov/gmd/ccgg/trends/ (accessed July 20, 2013).

United States Energy Information Administration. International Energy Statistics. http://www.eia.gov/cfapps/ipdbproject/IEDIndex3.cfm?tid=90&pid=44&aid=8 (accessed November 22, 2012).

United States Energy Information Administration. "International Energy Outlook 2010 – Highlights." 2010. http://www.eia.doe.gov/oiaf/ieo/highlights.html (accessed November 1, 2012).

United States Energy Information Administration. "Carbon Dioxide Emissions from the Generation of Electric Power in the United States." 2000. http://www.eia.doe .gov/cneaf/electricity/page/co2_report/co2report.html (accessed March 2, 2011).

University of California – Riverside. "Large Methane Release Could Cause Abrupt Climate Change as Happened 635 Million Years Ago," *ScienceDaily*, May 29, 2008. http://www.sciencedaily.com/releases/2008/05/080528140255.htm (accessed August 31, 2012).

Urban, Mark C., Kimberly S. Sheldon, and Josh J. Tewksbury. "On a Collision Course: Competition and Dispersal Differences Create No-Analogue Communities and Cause Extinctions during Climate Change," *Proceedings of the Royal Society B: Biological Sciences* 279 (2012): 2072–2080.

Vanderheiden, Steve. *Atmospheric Justice: A Political Theory of Climate Change*. Oxford: Oxford University Press, 2008.

Van Dyke, John C. *The Desert*. New York: Peregrine Smith, 1980.

Vattel, Emerich de. The Law of Nations, Or, Principles of the Law of Nature, Applied to the Conduct and Affairs of Nations and Sovereigns, with Three Early Essays on the Origin and Nature of Natural Law and on Luxury. 1797; The Online Library of Liberty, 2013. http://files.libertyfund.org/files/2246/Vattel_1519_EBk_v6.0.pdf (accessed December 18, 2013).

Victor, David G. *Global Warming Gridlock: Creating More Effective Strategies for Protecting the Planet*. Cambridge: Cambridge University Press, 2011.

Waldron, Jeremy. "Can Communal Goods be Human Rights?" in *Liberal Rights: Collected Papers 1981–1991*. Cambridge: Cambridge University Press, 1993.

Warner, John T. and Saul Pleeter. "The Personal Discount Rate," *American Economic Review* 91 (2001): 33–53.

Warren, Rachel. "The Role of Interactions in a World Implementing Adaptation and Mitigation Solutions to Climate Change," *Philosophical Transactions of the Royal Society A* 369 (2001): 217–241.

Weisbach, David. *Negligence, Strict Liability, and Responsibility for Climate Change*. The Harvard Project on International Climate Agreements: 2010. http://live .belfercenter.org/files/WeisbachDP39.pdf (accessed June 27, 2011).

Weitzman, Martin L. "A Review of the Stern Review on the Economics of Climate Change." *Journal of Economic Literature* 45 (2007): 703–724.

Wigely, T. M. L. "The Climate Change Commitment," *Science* 307, no. 5716 (2005): 1766–1769.

Williams, Bernard. "A Critique of Utilitarianism," in *Utilitarianism: For and Against*, by Bernard Williams and J. J. C. Smart. Cambridge: Cambridge University Press, 1973.

Wilson, Eward O. *The Diversity of Life*. Cambridge, MA: Harvard University Press, 2010.

Woodward, James. "The Non-Identity Problem," *Ethics* 96 (1986): 804–831.

World Health Organization. *Submission to the Ad Hoc Working Group on Long-Term Cooperative Action*. Bonn: World Health Organization, 2009. http://www.unfccc .int/resource/docs/2009/smsn/igo/047.pdf (accessed April 29, 2013).

Young, Iris Marion. *Responsibility for Justice*. Oxford: Oxford University Press, 2001.

Index